Geometric Programming for Design and Cost Optimization

(with illustrative case study problems and solutions)

Second Edition

Copyright © 2011 by Morgan & Claypool

All rights reserved. No part of this publication may be reproduced, stored in a retrieval system, or transmitted in any form or by any means—electronic, mechanical, photocopy, recording, or any other except for brief quotations in printed reviews, without the prior permission of the publisher.

Geometric Programming for Design and Cost Optimization (with illustrative case study problems and solutions) - Second Edition

Robert C. Creese

www.morganclaypool.com

ISBN: 9781608456109   paperback
ISBN: 9781608456116   ebook

DOI 10.2200/S00314ED1V01Y201011ENG012

A Publication in the Morgan & Claypool Publishers series
SYNTHESIS LECTURES ON ENGINEERING

Lecture #12
Series ISSN
Synthesis Lectures on Engineering
Print 1939-5221   Electronic 1939-523X

Synthesis Lectures on Engineering

Geometric Programming for Design and Cost Optimization (with illustrative case study problems and solutions) - Second Edition
Robert C. Creese
2010

Survive and Thrive: A Guide for Untenured Faculty
Wendy C. Crone
August 2010

Geometric Programming for Design and Cost Optimization (with Illustrative Case Study Problems and Solutions)
Robert C. Creese
2009

Style and Ethics of Communication in Science and Engineering
Jay D. Humphrey and Jeffrey W. Holmes
2008

Introduction to Engineering: A Starter's Guide with Hands-On Analog Multimedia Explorations
Lina J. Karam and Naji Mounsef
2008

Introduction to Engineering: A Starter's Guide with Hands-On Digital Multimedia and Robotics Explorations
Lina J. Karam and Naji Mounsef
2008

CAD/CAM of Sculptured Surfaces on Multi-Axis NC Machine: The DG/K-Based Approach
Stephen P. Radzevich
2008

Tensor Properties of Solids, Part Two: Transport Properties of Solids
Richard F. Tinder
2007

Tensor Properties of Solids, Part One: Equilibrium Tensor Properties of Solids
Richard F. Tinder
2007

Essentials of Applied Mathematics for Scientists and Engineers
Robert G. Watts
2007

Project Management for Engineering Design
Charles Lessard and Joseph Lessard
2007

Relativistic Flight Mechanics and Space Travel
Richard F. Tinder
2006

Geometric Programming for Design and Cost Optimization

(with illustrative case study problems and solutions)

Second Edition

Robert C. Creese
West Virginia University

SYNTHESIS LECTURES ON ENGINEERING #12

ABSTRACT

Geometric programming is used for design and cost optimization, the development of generalized design relationships, cost ratios for specific problems, and profit maximization. The early pioneers of the process - Zener, Duffin, Peterson, Beightler, Wilde, and Phillips – played important roles in the development of geometric programming. There are three major areas: 1) Introduction, History, and Theoretical Fundamentals, 2) Applications with Zero Degrees of Difficulty, and 3) Applications with Positive Degrees of Difficulty. The primal-dual relationships are used to illustrate how to determine the primal variables from the dual solution and how to determine additional dual equations when the degrees of difficulty are positive. A new technique for determining additional equations for the dual, Dimensional Analysis, is demonstrated. The various solution techniques of the constrained derivative approach, the condensation of terms, and dimensional analysis are illustrated with example problems. The goal of this work is to have readers develop more case studies to further the application of this exciting tool.

KEYWORDS

posynominials, primal, dual, cost optimization, design optimization, generalized design relationships, cost ratios, profit maximization, constrained derivatives, dimensional analysis, condensation of terms

Contents

Preface xi

PART I Introduction, History, and Theoretical Fundamentals of Geometric Programming 1

1 Introduction 3
1.1 Optimization and Geometric Programming 3
    1.1.1 Optimization 3
    1.1.2 Geometric Programming 3
1.2 Evaluative Questions 4
References 4

2 Brief History of Geometric Programming 5
2.1 Pioneers of Geometric Programming 5
2.2 Evaluative Questions 6
References 6

3 Theoretical Considerations 7
3.1 Primal and Dual Formulation 7
3.2 Evaluative Questions 10
References 10

PART II Geometric Programming Applications with Zero Degrees of Difficulty 11

4 The Optimal Box Design Case Study 13
4.1 The Optimal Box Design Problem 13
4.2 Evaluative Questions 16

5 Trash Can Case Study ... 17
- 5.1 Introduction ... 17
- 5.2 Problem Statement and General Solution ... 17
  - 5.2.1 Example ... 17
- 5.3 Evaluative Questions ... 20

6 The Open Cargo Shipping Box Case Study ... 21
- 6.1 Problem Statement and General Solution ... 21
- 6.2 Evaluative Questions ... 24
  - References ... 25

7 Metal Casting Cylindrical Riser Case Study ... 27
- 7.1 Introduction ... 27
- 7.2 Problem Formulation and General Solution ... 29
- 7.3 Example ... 30
- 7.4 Evaluative Questions ... 31
  - References ... 31

8 Inventory Model Case Study ... 33
- 8.1 Problem Statement and General Solution ... 33
- 8.2 Example ... 35
- 8.3 Evaluative Questions ... 36
  - References ... 36

9 Process Furnace Design Case Study ... 37
- 9.1 Problem Statement and Solution ... 37
- 9.2 Evaluative Questions ... 41
  - References ... 41

10 Gas Transmission Pipeline Case Study ... 43
- 10.1 Problem Statement and Solution ... 43
- 10.2 Evaluative Questions ... 46
  - References ... 46

11	**Profit Maximization Case Study** 47	
	11.1 Profit Maximization and Geometric Programming 47	
	11.2 Profit Maximization using the Cobb-Douglas Production Function 47	
	11.3 Evaluative Questions 49	
	References 50	
12	**Material Removal/Metal Cutting Economics Case Study** 51	
	12.1 Introduction 51	
	12.2 Problem Formulation 51	
	12.3 Evaluative Questions 54	
	References 55	

PART III Geometric Programming Applications with Positive Degrees of Difficulty 57

13	**Journal Bearing Design Case Study** 59	
	13.1 Introduction 59	
	13.2 Primal and Dual Formulation of Journal Bearing Design 59	
	13.3 Dimensional Analysis Technique for Additional Equation 63	
	13.4 Evaluative Questions 64	
	References 65	
14	**Metal Casting Hemispherical Top Cylindrical Side Riser Case Study** 67	
	14.1 Introduction 67	
	14.2 Problem Formulation 67	
	14.3 Dimensional Analysis Technique for Additional Two Equations 72	
	14.4 Evaluative Questions 75	
	References 75	
15	**Liquefied Petroleum Gas (LPG) Cylinders Case Study** 77	
	15.1 Introduction 77	
	15.2 Problem Formulation 77	
	15.3 Dimensional Analysis Technique for Additional Equation 82	
	15.4 Evaluative Questions 83	
	References 84	

16 Material Removal/Metal Cutting Economics with Two Constraints 85
- 16.1 Introduction 85
- 16.2 Problem Formulation 85
- 16.3 Problem Solution 87
- 16.4 Example Problem 91
- 16.5 Evaluative Questions 91
  - References 92

17 The Open Cargo Shipping Box with Skids 95
- 17.1 Introduction 95
- 17.2 Primal-Dual Problem Formulation 95
- 17.3 Constrained Derivative Approach 97
- 17.4 Dimensional Analysis Approach for Additional Equation 98
- 17.5 Condensation of Terms Approach 100
- 17.6 Evaluative Questions 102
  - References 102

18 Profit Maximization Considering Decreasing Cost Functions of Inventory Policy 103
- 18.1 Introduction 103
- 18.2 Model Formulation 103
- 18.3 Example 107
- 18.4 Transformed Dual Approach 109
- 18.5 Evaluative Questions 112
  - References 112

19 Summary and Future Directions 115
- 19.1 Summary 115
- 19.2 Future Directions 115
- 19.3 Development of New Design Relationships 115

20 Thesis and Dissertations on Geometric Programming 119

Author's Biography 125

Index 127

Preface

The purpose of this text is to introduce manufacturing engineers, design engineers, manufacturing technologists, cost engineers, project managers, industrial consultants and finance managers to the topic of geometric programming. I was fascinated by the topic when first introduced to it at a National Science Foundation (NSF) Short Course in Austin, Texas in 1967. The topic was only for a day or so of a three week course, but I recognized its potential in the application to riser design in the metal casting industry during the presentation. I was fortunate to have two of the pioneers in Geometric Programming make the presentations, Doug Wilde of Stanford University and Chuck Beightler of the University of Texas, and had them autograph their book "Foundations of Optimization" for me which I fondly cherish even now.

I finally wrote my first publication using geometric programming in 1972 and have written several journal papers using geometric programming on metal cutting and metal casting riser design problems, but was able to teach a complete course on the topic. Thus, before I retire, I decided to write a brief book on the topic illustrating the basic approach to solving various problems to encourage others to pursue the topic in more depth. Its ability to lead to design and cost relationships in an integrated manner makes this tool essential for engineers, product developers and project managers be more cost competitive in this global market place.

This book is dedicated to the pioneers of geometric programming such as Clarence Zener, Richard Duffin, Elmor Peterson, Chuck Beightler, Doug Wilde, Don Phillips, and several others for developing this topic. This work is also dedicated to my family members, Natalie and Jennifer; Rob, Denie, Robby and Sammy, and Chal and Joyce.

I also want to recognize those who have assisted me in reviewing and editing of this work and they are Dr. M. Adithan, Senior Professor in Mechanical Engineering and Dean of Faculty at VIT University, Vellore, Tamil Nadu, India and Dr. Deepak Gupta, Assistant Professor at Southeast Missouri State University, USA.

I used the first edition of the book to teach a course on geometric programming at the MS level. The students, Yi Fang, Mohita Yalamanchi, Srikanth Manukonda, Shri Harsha Chintala, Kartik Ramamoorthy, and Joshua Billups developed various examples, both original and from the literature which are included in this new edition and they taught me a lot about geometric programming. One of the important items presented during the course was the dimensional analysis technique for the obtaining of additional equations when the degrees of difficulty was positive, and no reference to this technique could be found in the literature.

December 2010

Dr. Robert C. Creese, CCE
Industrial & Management Systems Engineering
West Virginia University, Morgantown, West Virginia, USA

PART I

Introduction, History, and Theoretical Fundamentals of Geometric Programming

CHAPTER 1

Introduction

1.1 OPTIMIZATION AND GEOMETRIC PROGRAMMING

1.1.1 OPTIMIZATION

Optimization can be defined as the process of determining the best or most effective result utilizing a quantitative measurement system. The measurement unit most commonly used in financial analysis, engineering economics, cost engineering or cost estimating tends to be currency such as US Dollars, Euros, Rupees, Yen, Won, Pounds Sterling, Kroner, Kronor, Pesos or specific country currency. The optimization may occur in terms of net cash flows, profits, costs, benefit/cost ratio etc. Other measurement units may be used, such as units of production or production time, and optimization may occur in terms of maximizing production units, minimizing production time, maximizing profits, or minimizing cost. Design optimization determines the best design that meets the desired design constraints at the desired objective, which typically is the minimum cost. Two of the most important criteria for a successful product are to meet all the functional design requirements and to be economically competitive.

There are numerous techniques of optimization methods such as linear programming, dynamic programming, geometric programming, queuing theory, statistical analysis, risk analysis, Monte Carlo simulation, numerous search techniques, etc. Geometric programming is one of the better tools that can be used to achieve the design requirements and minimal cost objective. The development of the concept of geometric programming started in 1961. Geometric programming can be used not only to provide a specific solution to a problem, but it also can in many instances give a general solution with specific design relationships. These design relationships, based upon the design constants, can then be used for the optimal solution without having to resolve the original problem. A second concept is that the dual solution gives a constant ratio between the terms of the objective function. These fascinating characteristics appear to be unique to geometric programming.

1.1.2 GEOMETRIC PROGRAMMING

Geometric programming is a mathematical technique for optimizing positive polynomials, which are called posynominials. This technique has many similarities to linear programming, but has advantages in that:

1. a non-linear objective function can be used;

2. the constraints can be non-linear; and

3. the optimal cost value can be determined with the dual without first determining the specific values of the primal variables.

Geometric programming can lead to generalized design solutions and specific relationships between variables. Thus, a cost relationship can be determined in generalized terms when the degrees of difficulty are low, such as zero or one. This major disadvantage is that the mathematical formulation is much more complex than linear programming and complex problems are very difficult to solve. It is called geometric programming because it is based upon the arithmetic-geometric inequality where the arithmetic mean is always greater than or equal to the geometric mean. That is:

$$(X_1 + X_2 + \ldots X_n)/n \geq (X_1 * X_2 * \ldots * X_n)^{(1/n)} . \tag{1.1}$$

Geometric programming was first presented over 50 years but has not received adequate attention similar to that which linear programming has obtained over its history of less than 70 years. Some of the early historical highlights and achievements of geometric programming developments are presented in the next chapter.

1.2   EVALUATIVE QUESTIONS

1. What is the most common unit of measurement used for optimization?

2. The following series of costs ($) were collected: 2, 4, 6, 8, and 10

    (a) What is the arithmetic mean of the series of costs?

    (b) What is the geometric mean of the series of costs?

3. The following series of costs (€) were collected: 20, 50, 100, 500, and 600

    (a) What is the arithmetic mean of the series of costs?

    (b) What is the geometric mean of the series of costs?

4. The following series of costs (Rupees) were collected: 5, 7, 8, 12, 16, and 18

    (a) What is the arithmetic mean of the series of costs?

    (b) What is the geometric mean of the series of costs?

5. What is the year recognized as the beginning of geometric programming?

REFERENCES

[1] R.J. Duffin, E.L. Peterson and C. Zener, *Geometric Programming*, John Wiley and Sons, New York, 1967.

CHAPTER 2

Brief History of Geometric Programming

2.1 PIONEERS OF GEOMETRIC PROGRAMMING

Clarence Zener, Director of Science at Westinghouse Electric in Pittsburgh, Pennsylvania, USA, is credited as being the father of Geometric Programming. In 1961 he published a paper in the Proceedings of the National Academy of Science on "A mathematical aid in optimizing engineering designs" [6] which is considered as the first paper on geometric programming. Clarence Zener is better known in electrical engineering for the Zener diode. He later teamed with Richard J. Duffin and Elmor L. Peterson of the Carnegie Institute of Technology (now Carnegie-Mellon University, USA) to write the first book on geometric programming, named "Geometric Programming" in 1967 [1]. A report by Professor Douglas Wilde and graduate student Ury Passey on "Generalized Polynomial Optimization" was published in August 1966. Professor Douglas Wilde of Stanford University and Professor Charles Beightler of the University of Texas included a chapter on Geometric Programming in their text "Foundations of Optimization" [2]. I attended an Optimization Short Course at the University of Texas in August 1967 and that is when I first became interested in geometric programming. I realized at that time that geometric programming could be used for the metal casting riser design problem, and I published a paper on it in 1971 [7].

Other early books by these leaders were "Engineering Design by Geometric Programming" by Clarence Zener in 1971 [3], "Applied Geometric Programming" by C.S. Beightler and D.T. Phillips in 1976, and the second edition of "Foundations of Optimization" by C.S. Beightler, D.T. Phillips, and D. Wilde in 1979 [4]. Many of the initial applications were in the area of transformer design as Clarence Zener worked for Westinghouse Electric and in the area of Chemical Engineering which was the area emphasized by Beightler and Wilde. It is also important that several graduate students played an important role in the development of geometric programming, namely Elmor Peterson at Carnegie Institute of Technology and Ury Passy and Mordecai Avriel at Stanford University.

Geometric programming has attracted a fair amount of interest and a list of the various thesis and dissertations published that have either mentioned or focused on geometric programming are listed in an appendix. Web sites on geometric programming [5] have appeared with additional interesting applications.

2.2 EVALUATIVE QUESTIONS

1. Who is recognized as the father of geometric programming?

2. When was the first book published on geometric programming and what was the title of the book?

3. Which three Universities played an important role in the development of geometric programming?

REFERENCES

[1] R.J. Duffin, E.L. Peterson and C. Zener, *Geometric Programming*, John Wiley and Sons, New York, 1967. 5

[2] D.J. Wilde and C.S. Beighler, *Foundations of Optimization*, Prentice-Hall, Englewood Cliffs, New Jersey, 1967. 5

[3] C. Zener, *Engineering Design by Geometric Programming*, John Wiley and Sons, New York, 1971. 5

[4] C.S. Beightler, D.T. Phillips and D.J. Wilde, *Foundations of Optimization*, 2nd Edition, Prentice Hall, Englewood Cliffs, New Jersey, 1979. 5

[5] http://www.mpri.lsu.edu/textbook/Chapter3.htm (Chapter 3 Geometric Programming) (site visited 5–20-09) 5

[6] C. Zener, "A Mathematical Aid in Optimizing Engineering Design", *Proceedings of the National Academy of Science*, Vol. 47, 1961, p. 537. 5

[7] R.C. Creese, "Optimal Riser Design by Geometric Programming", *AFS Cast Metals Research Journal*, Vol. 7, 1971, pp. 118–121. 5

CHAPTER 3

Theoretical Considerations

3.1 PRIMAL AND DUAL FORMULATION

Geometric programming requires that the expressions used are posynomials, and it is necessary to distinguish between functions, monomials and posynomials [1]. Posynomial is meant to indicate a combination of "positive" and "polynomial" and implies a "positive polynomial." Examples of functions, which are monomials, are:

$$5x, \quad 0.25, \quad 4x^2, \quad 2x^{1.5}y^{-.15}, \quad 160, \quad 65x^{-15}t^{10}z^2.$$

Examples of posynomials are, which are monomials or sums of monomials, are:

$$5 + xy, \quad (x + 2YZ)^2, \quad x + 2y + 3z + t \quad x/y + z35x^{1.5} + 72Y^3.$$

Examples of expressions which are not posynomials are:

-1.5 (negative sign)

$(2 + 2yz)^{3.2}$ (fractional power of multiple term which cannot be expanded)

$x - 2y + 3z$ (negative sign)

$x + \sin(x)$ (sine expression can be negative).

The coefficients of the constants must be positive, but the coefficients of the exponents can be negative.

The mathematics of geometric programming are rather complex, however the basic equations are presented and followed by an illustrative example. The theory of geometric programming is presented in more detail in some of the references [1, 2, 3, 4, 5] listed at the end of the chapter. The primal problem is complex, but the dual version is much simpler to solve. The dual is the version typically solved, but the relationships between the primal and dual are needed to determine the specific values of the variables in the primal. The primal problem is formulated as:

$$Y_m(X) = \sum_{T=1}^{T_m} \sigma_{mt} C_{mt} \prod_{n=1}^{N} X_n^{a_{mtn}} ; \quad m = 0, 1, 2, \ldots M , \tag{3.1}$$

3. THEORETICAL CONSIDERATIONS

with $\quad \sigma_{mt} = \pm 1$ and $C_{mt} > 0$
and $\quad Y_m(X) \leq \sigma_m$ for $m = 1, \ldots M$ for the constraints
where $\quad C_{mt}$ = positive constant coefficients in cost and constraint equations
and $\quad Y_m(X)$ = primal objective function
and $\quad \sigma_{mt}$ = signum function used to indicate sign of term in the equation (either $+1$ or -1).

The dual is the problem formulation that is typically solved to determine the dual variables and value of the objective function. The dual objective function is expressed as:

$$d(\omega) = \sigma \left[\prod_{m=0}^{M} \prod_{t=1}^{T_m} (C_{mt}\omega_{m0}/\omega_{mt})^{\sigma_{mt}\omega_{mt}} \right]^{\sigma} \quad m = 0, 1, \ldots M \text{ and } t = 1, 2, \ldots T_m, \quad (3.2)$$

where
$\quad \sigma \quad =$ signum function (± 1)
$\quad C_{mt} =$ constant coefficient
$\quad \omega_{m0} =$ dual variables from the linear inequality constraints
$\quad \omega_{mt} =$ dual variables of dual constraints, and
$\quad \sigma_{mt} =$ signum function for dual constraints,
and by definition:

$$\omega_{00} = 1. \quad (3.3)$$

The dual is formulated from four conditions:
(1) a normality condition

$$\sum_{T=1}^{T_m} \sigma_{0t}\omega_{0t} = \sigma \quad \text{where} \quad \sigma = \pm 1, \quad (3.4)$$

where
$\quad \sigma_{0t} =$ signum of objective function terms
$\quad \omega_{0t} =$ dual variables for objective function terms.
(2) N orthogonal conditions

$$\sum_{m=0}^{M} \sum_{t=1}^{T} \sigma_{mt} a_{mtn} \omega_{mt} = 0, \quad (3.5)$$

where
$\quad \sigma_{mt} =$ signum of constraint term
$\quad a_{mtn} =$ exponent of design variable term
$\quad \omega_{mt} =$ dual variable of dual constraint.
(3) T non-negativity conditions (dual variables must be positive):

$$\omega_{mt} \geq 0 \quad m = 0, 1 \ldots M \text{ and } t = 1, 2 \ldots \ldots T_m. \quad (3.6)$$

(4) M linear inequality constraints:

$$\omega_{mo} = \sigma_m \sum_{t=1}^{T_m} \sigma_{mt}\omega_{mt} \geq 0 \ . \tag{3.7}$$

The dual variables, ω_{mt}, are restricted to being positive, which is similar to the linear programming concept of all variables being positive. If the number of independent equations and variables in the dual are equal, the degrees of difficulty are zero. The degrees of difficulty is the difference between the number of dual variables and the number of independent linear equations; and the greater this degrees of difficulty, the more difficult the solution. The degrees of (D) can be expressed as:

$$D = T - (N+1) \tag{3.8}$$

where
- T = total number of terms (of primal)
- N = number of orthogonality conditions plus normality condition (which is equivalent to the number of primal variables).

Once the dual variables are found, the primal variables can be determined from the relationships:

$$C_{ot} \prod_{n=1}^{N} X_n^{a_{mtn}} = \omega_{ot}\sigma Y_o \qquad t = 1, \ldots T_o \ , \tag{3.9}$$

and

$$C_{mt} \prod_{n=1}^{N} X_n^{a_{mtn}} = \omega_{mt}/\omega_{mo} \qquad t = 1, \ldots T_o \text{ and } m = 1, \ldots M \ . \tag{3.10}$$

The theory may appear to be overwhelming with all the various terms, but various examples will be presented in the following chapters to illustrate the application of the various equations.

There are two sections of examples, the first considering basic examples with zero degrees of difficulty, and then the second section considering problems with more than zero degree of difficulty and presenting various approaches to solving the problem. Problems with zero degrees of difficulty typically will have a linear dual formulation with an equal number of equations and dual variables and can be solved relatively easily.

When there are more than zero degrees of difficulty, the solution is much more difficult. Some of the approaches to solve these problems are:

1) Finding additional equations so the number of variables and number of equations are equal, but the additional equations are usually non-linear. The additional equations can be determined by either: a) dimensional analysis of the primal dual relationships or b) by substitution techniques.

2) Express the dual in terms of only one dual variable by substitution and take the derivative of the dual, set it to zero, and obtain the value of the unknown dual variable.

3) Condensation techniques by combining two or more of the primal terms to reduce the number of dual variables. This technique gives an approximate optimal solution and will be illustrated.

4) Transformed Problem Approach. This technique is often used on maximization problems to transform signomial problems into posynominial problems. It is also used to have the terms of the transformed objective function as a function of a single variable to make the solution easier.

3.2 EVALUATIVE QUESTIONS

1. What version of the geometric problem formulation is solved for the objective function and why?

2. What values can the signum function have?

3. How are the primal variables determined?

4. Which of the following terms are posynomials?

   a) 3.4 b) $4x$ c) $5xy$ d) $4x^{2.1}$ e) $(x+2)^4$
   f) $5x - 3$ g) $6x^{-2.4}$ h) e^{-3} i) $3x + 4y + 5z^{-1.4}$ j) $(x+4)^{2.2}$
   k) $(x - y + 3)^2$ l) $\cot y$ m) $x/2y$ n) $5l^{-2}t^5$

REFERENCES

[1] S. Boyd, S-J Kim, L. Vandengerghe, and A. Hassibi, "A Tutorial on Geometric Programming", *Optimization and Engineering*, 8(1), pp 67–127, Springer, Germany, 2007. 7

[2] R.J. Duffin, E.L. Peterson and C. Zener, *Geometric Programming*, John Wiley and Sons, New York, 1967. 7

[3] D.J. Wilde and C.S. Beighler, *Foundations of Optimization*, Prentice-Hall, Englewood Cliffs, New Jersey, 1967. 7

[4] C. Zener, *Engineering Design by Geometric Programming*, John Wiley and Sons, New York, 1971. 7

[5] C.S. Beightler, D.T. Phillips and D.J. Wilde, *Foundations of Optimization*, 2nd Edition, Prentice Hall, Englewood Cliffs, New Jersey, 1979. 7

PART II

Geometric Programming Applications with Zero Degrees of Difficulty

CHAPTER 4

The Optimal Box Design Case Study

4.1 THE OPTIMAL BOX DESIGN PROBLEM

The optimal box design problem is a relatively easy problem which illustrates the procedure for solving a geometric programming problem with zero degrees of difficulty. It also indicates the importance of a general solution which is possible with geometric programming; that is formulas for the box dimensions can be developed which will give the answers without needing to resolve the problem if the costs or box volume changes.

EXAMPLE:

A box manufacturer wants to determine the optimal dimensions for making boxes to sell to customers. The cost for production of the sides is C_1 ($ 2/sq ft) and the cost for producing the top and bottom is C_2 ($ 3/sq ft) as more cardboard is used for the top and bottom of the boxes. The volume of the box is to be set at a limit of "V" (4 ft^3) which can be varied for different customer specifications. If the dimensions of the box are W for the width, H for the box height, and L for the box length, what should the dimensions be based upon the cost values and box volume?

The problem is to minimize the box cost for a specific box volume. The primal objective function is:

$$\text{Minimize} \quad \text{Cost}(Y) = C_2 W L + C_1 H (W + L) \quad (4.1)$$
$$\text{Subject to:} \quad W L H \geq V . \quad (4.2)$$

However, in geometric programming the inequalities must be written in the form of $\leq$ and the right-hand side must be ± 1. Thus, the primal constraint becomes:

$$\text{Minimize} \quad \text{Cost}(Y) = C_1 H W + C_1 H L + C_2 W L \quad (4.3)$$
$$\text{Subject to} \quad -W H L / V \leq -1 . \quad (4.4)$$

4. THE OPTIMAL BOX DESIGN CASE STUDY

From the coefficients and signs, the signum values for the dual are:

$$\sigma_{01} = 1$$
$$\sigma_{02} = 1$$
$$\sigma_{03} = 1$$
$$\sigma_{11} = -1$$
$$\sigma_{1} = -1$$

Thus, the dual formulation from Chapter 3 would be:

Objective Function (using Eqs. (3.4) and (4.3))	$\omega_{01} + \omega_{02} + \omega_{03}$	$= 1$	(4.5)
L terms (using Eqs. (3.5), (4.3), and (4.4))	$\omega_{02} + \omega_{03} - \omega_{11}$	$= 0$	(4.6)
H terms (using Eqs. (3.5), (4.3), and (4.4))	$\omega_{01} + \omega_{02} - \omega_{11}$	$= 0$	(4.7)
W terms (using Eqs. (3.5), (4.3), and (4.4))	$\omega_{01} + \omega_{03} - \omega_{11}$	$= 0$	(4.8)

The degrees of difficulty (D) are equal to:

$$D = T - (N + 1) = 4 - (3 + 1) = 0,$$

where
$T = $ total number of terms of primal and
$N = $ number of orthogonality conditions plus normality condition or the number of primal variables (H, W, and L gives 3 primal variables).

Thus, one has the same number of variables as equations, so this can be solved by simultaneous equations as these are linear equations.

Using Equations (4.5)–(4.8), the values for the dual variables are found to be:

$$\omega_{01} = 1/3$$
$$\omega_{02} = 1/3$$
$$\omega_{03} = 1/3$$
$$\omega_{11} = 2/3$$

and by definition

$$\omega_{00} = 1$$

Using the linearity inequality equation expressed by Equation (3.7),

$$\omega_{10} = \omega_{mt} = \sigma_m \sum \sigma_{mt} \omega_{mt} = (-1) * (-1 * 2/3) = 2/3 > 0 \text{ where } m = 1 \text{ and } t = 1.$$

4.1. THE OPTIMAL BOX DESIGN PROBLEM

The objective function can be found using Equation (3.2):

$$d(\omega) = \sigma \left[\prod_{m=0}^{M} \prod_{t=1}^{T_m} (C_{mt}\omega_{mo}/\omega_{mt})^{\sigma_{mt}\omega_{mt}} \right]^{\sigma} \qquad (3.2)$$

$$d(\omega) = 1\Big[\{(C_1 * 1)/(1/3)\}^{(1)*(1/3)} * \{(C_1 * 1)/(1/3)\}^{(1)*(1/3)}$$
$$* \{(C_2 * 1)/(1/3)\}^{(1)*(1/3)} * \{((1/V) * 1)/(2/3)\}^{(-1)*(2/3)}\Big]^1$$
$$= 1\Big[\{(3C_1)^{1/3}\} * \{(3C_1)^{1/3}\} * \{(3C_2)^{1/3}\} * \{(1/V)^{-2/3}\}\Big]$$
$$= 3C_1^{2/3} C_2^{1/3} V^{2/3} \qquad (4.9)$$
$$= 3\left(2^{2/3} 3^{1/3} 4^{2/3}\right) = \$17.31 \ .$$

The solution has been determined without finding the values for L, W, or H. Also note that the dual expression is expressed in constants and thus the answer can be found without having to resolve the entire problem as one only needs to use the new constant values. To find the values of L, W, and H, one must use Equations (3.9) and (3.10), which are repeated here.

$$C_{ot} \prod_{n=1}^{N} X_n^{a_{mtn}} = \omega_{ot}\sigma Y_o \qquad t = 1, \ldots, T_o \qquad (3.9)$$

and

$$C_{mt} \prod_{n=1}^{N} X_n^{a_{mtn}} = \omega_{mt}/\omega_{mo} \qquad t = 1, \ldots, T_o \quad \text{and} \quad m = 1, \ldots, M \ . \qquad (3.10)$$

Using Equation (3.9) the relationships are:

$$C_1 HW = \omega_{01} Y = Y/3$$
$$C_1 HL = \omega_{02} Y = Y/3$$
$$C_2 WL = \omega_{03} Y = Y/3 \ .$$

Combining the first two of these relationships one obtains

$$W = L \ . \qquad (4.10)$$

Combining the last two of these relationships one obtains

$$H = (C_2/C_1)L \ . \qquad (4.11)$$

Since $V = HWL = (C_2/C_1)LLL = (C_2/C_1)L^3$.

4. THE OPTIMAL BOX DESIGN CASE STUDY

Thus,

$$L = [V(C_1/C_2)]^{1/3} \tag{4.12}$$
$$W = L = [V(C_1/C_2)]^{1/3} \tag{4.13}$$
$$H = (C_2/C_1)L = (C_2/C_1)[V(C_1/C_2)]^{1/3} = \left[V(C_2^2/C_1^2)\right]^{1/3}. \tag{4.14}$$

The specific values for this particular problem would be:

$$L = [4(2/3)]^{1/3} = 1.386 \text{ ft}$$
$$W = L = 1.386 \text{ ft}$$
$$H = \left[4(3^2/2^2)\right]^{1/3} = 2.080 \text{ ft}.$$

The volume of the box, $LWH = (1.386)(1.386)(2.080) = 4.0 \text{ ft}^3$, which was the minimum volume required for the box. To verify the results, the parameters are used in the primal problem (Equation (4.3)) to make certain the solution obtained is the same.

$$\text{Cost}(Y) = C_1 HW + C_1 HL + C_2 WL \tag{4.3}$$

$$Y_0 = C_1 \left[V\left(C_2^2/C_1^2\right)\right]^{1/3} [V(C_1/C_2)]^{1/3} + C_1 \left[V\left(C_2^2/C_1^2\right)\right]^{1/3} [V(C_1/C_2)]^{1/3}$$
$$+ C_2 [V(C_1/C_2)]^{1/3} [V(C_1/C_2)]^{1/3}$$
$$Y_0 = C_1^{2/3}\left(V^{2/3}\right) C_2^{1/3} + C_1^{2/3}\left(V^{2/3}\right) C_2^{1/3} + C_1^{2/3}\left(V^{2/3}\right) C_2^{1/3}$$
$$Y_0 = 3 C_1^{2/3} C_2^{1/3} V^{2/3}. \tag{4.15}$$

The expressions for Equation (4.9) from the primal and Equation (4.15) from the dual are equivalent. The geometric programming solution is in general terms, and thus can be used for any values of C_1, C_2, and V. This ability to obtain general relationships makes the use of Geometric Programming a very valuable tool for cost engineers.

4.2 EVALUATIVE QUESTIONS

1. A large box is to be made with the values of $C_1 = 4$ Euros/m², $C_2 = 4$ Euros/m², and $V = 8$ m³. What is the cost (Euros) and the values of H, W, and L?

2. The cost of the top and bottom is increased to 6 Euros/m² and what is the increase in the box cost and the change in box dimensions?

3. The box is to be open (that is there is no top). Determine the expressions for H, W, and L for an open box and determine the cost if $C_1 = 4$ Euros/m², $C_2 = 6$ Euros/m², and $V = 8$ m³. Compare the results with Problem 2 and discuss the differences in the formulas, dimensions, and costs.

CHAPTER 5
Trash Can Case Study

5.1 INTRODUCTION

Various case studies are used to illustrate the different applications of geometric programming as well as to illustrate the different conditions that must be evaluated in solving the problems. The second case study, the trash can case study, is easy to solve and has zero degrees of difficulty. It is similar to the box problem, but involves a different shape. The solution is provided in detail, giving the general solution for the problem in addition to the specific solution. These examples are provided so that the readers can develop solutions to specific problems that they may have and to illustrate the importance of the generalized solution.

5.2 PROBLEM STATEMENT AND GENERAL SOLUTION

5.2.1 EXAMPLE

Bjorn of Sweden has entered into the trash can manufacturing business and he is making cylindrical trash cans and wants to minimize the material cost. The trash can is an open cylinder and designed to have a specific volume. The objective will be to minimize the total material cost of the can. Figure 5.1 is a sketch of the trash can illustrating the design parameters used, the radius and the height of the trash can. The bottom and sides can be of different costs as the bottom is typically made of a thicker material. The primal objective function is:

$$\text{Minimize:} \quad \text{Cost}(Y) = C_1 \pi r^2 + C_2 2\pi r h \quad (5.1)$$
$$\text{Subject to:} \quad V = \pi r^2 h, \quad (5.2)$$

where:
- r = radius of trash can bottom
- h = height of trash can
- V = volume of trash can
- C_1 = material constant cost of bottom material of trash can
- C_2 = material constant cost of side material of trash can.

The constraint must be written in the form of an inequality, so

$$V \geq \pi r^2 h. \quad (5.3)$$

And it must be written in the less than equal form, so it becomes

$$-\pi r^2 h / V \leq -1. \quad (5.4)$$

5. TRASH CAN CASE STUDY

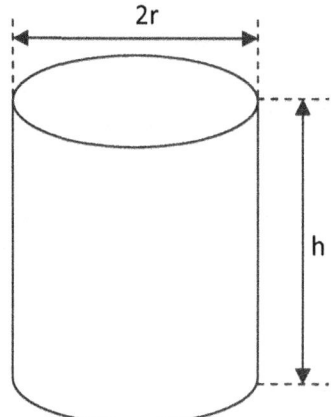

Figure 5.1: Trash can parameters r and h.

Thus, the primal problem is:

$$\text{Minimize:} \quad \text{Cost}(Y) = C_1 \pi r^2 + C_2 2\pi r h \quad (5.5)$$
$$\text{Subject to:} \quad -\pi r^2 h / V \leq -1. \quad (5.6)$$

From the coefficients and signs, the signum values for the dual are:

$$\sigma_{01} = 1$$
$$\sigma_{02} = 1$$
$$\sigma_{11} = -1$$
$$\sigma_1 = -1.$$

The dual formulation is:

$$\text{Objective Function} \quad \omega_{01} + \omega_{02} = 1 \quad (5.7)$$
$$r \text{ terms} \quad 2\omega_{01} + \omega_{02} - 2\omega_{11} = 0 \quad (5.8)$$
$$h \text{ terms} \quad \omega_{02} - \omega_{11} = 0. \quad (5.9)$$

Using theses equations, the values of the dual variables are found to be:

$$\omega_{01} = 1/3$$
$$\omega_{02} = 2/3$$
$$\omega_{11} = 2/3$$

and by definition

$$\omega_{00} = 1$$

The degrees of difficulty are equal to:
$$D = T - (N+1) = 3 - (2+1) = 0. \tag{5.10}$$

Using the linearity inequality equation,
$$\omega_{10} = \omega_{mt} = \sigma_m \sum \sigma_{mt}\omega_{mt} = (-1)*(-1*2/3) = 2/3 > 0 \quad \text{where } m=1 \text{ and } t=1.$$

The objective function can be found using the dual expression:
$$Y = d(\omega) = \sigma \left[\prod_{m=0}^{M} \prod_{t=1}^{T_m} (C_{mt}\,\omega_{mo}/\omega_{mt})^{\sigma_{mt}\omega_{mt}} \right]^\sigma \tag{5.11}$$
$$= 1\left[\left[\{(\pi\,C_1 * 1/(1/3))\}^{(1*1/3)} \right] \left[\{\pi\,C_2 * 1/(2/3))\}^{(1*2/3)} \right] \left[\{(\pi/V)*((2/3)/(2/3)\}^{(-1*2/3)} \right] \right]^1$$
$$= 3\pi^{1/3} C_1^{1/3} C_2^{2/3} V^{2/3}. \tag{5.12}$$

The values for the primal variables can be determined from the relationships between the primal and dual as:
$$C_1 \pi r^2 = \omega_{01} Y = 1/3 * Y. \tag{5.13}$$

And
$$C_2 2\pi r h = \omega_{02} Y = 2/3 * Y. \tag{5.14}$$

Dividing these expressions and reducing terms one can obtain:
$$r = (C_2/C_1) * h. \tag{5.15}$$

Setting
$$V = \pi r^2 h. \tag{5.16}$$

And using the last two equations one can obtain
$$h = ((V/\pi)(C_1^2/C_2^2))^{1/3}. \tag{5.17}$$

and
$$r = ((V/\pi)(C_2/C_1))^{1/3}. \tag{5.18}$$

Using (5.17) and (5.18) in Equation (5.5) for the primal, one obtains:
$$Y = C_1 \pi r^2 + C_2 2\pi r h \tag{5.5}$$
$$= 3\pi^{1/3} C_1^{1/3} C_2^{2/3} V^{1/3}. \tag{5.19}$$

5. TRASH CAN CASE STUDY

Note that Equations (5.19) and (5.12) are identical, which is what should happen, as the primal and dual objective functions must be identical.

An important aspect about the dual variables is that they indicate the effect of the terms upon the solution. The values of $\omega_{02} = 2/3$ and $\omega_{01} = 1/3$ indicates that the second term has twice the impact as the first term in the primal. For example, if $C_1 = 9$ \$/sq ft, $C_2 = 16$ \$/sq ft, and $V = 4\pi = 12.57$ cubic feet, then

$$h = ((V/\pi)(C_1^2/C_2^2))^{1/3} = ((4\pi/\pi)(9^2/16^2))^{1/3} = (4*81/256)^{1/3} = 1.082 \text{ ft}$$
$$\text{and} \quad r = ((V/\pi)(C_2/C_1))^{1/3} = ((4\pi/\pi)(16/9))^{1/3} = (4*16/9)^{1/3} = 1.923 \text{ ft}.$$

Note that:

$$V = \pi r^2 h = 3.1416 * 1.082 \text{ ft} * (1.923 \text{ ft})^2 = 12.57 \text{ ft}^3$$
$$\text{And} \quad Y = C_1 \pi r^2 + C_2 2\pi r h = 9 * 3.14 * 1.923^2 + 16 * 2 * 3.14 * 1.923 * 1.082$$
$$= \$104.5 + \$209.0$$
$$= \$313.5.$$

The contribution of the second term is twice that of the first term which is what is predicted by the value of the dual variables. This occurs regardless of the values of the constants used and this is an important concept for cost analysis.

5.3 EVALUATIVE QUESTIONS

1. A trash can is designed to hold 3 cubic meters of trash. Determine the cost and the design parameters (radius and height) in meters for if the costs C_1 and C_2 are 20 Swedish Kroner per square meter and 10 Swedish Kroner per square meter, respectively.

2. If the volume is doubled to 6 cubic meters, what are the new dimensions and cost?

3. If the trash can is to have a lid which will have the same diameter as the bottom of the trash can, what are the cost and dimensions of the trash can with the lid?

CHAPTER 6

The Open Cargo Shipping Box Case Study

6.1 PROBLEM STATEMENT AND GENERAL SOLUTION

This is a classic geometric programming problem as it was the first illustrative problem presented in the first book [1] on geometric programming. This problem presented here is expanded as not only is the minimum total cost required, but also the dimensions of the box. The problem is: "Suppose that 400 cubic yards (V) of gravel must be ferried across a river. The gravel is to be shipped in an open cargo box of length x_1, width x_2 and height x_3. The sides and bottom of the box cost $ 10 per square yard (A_1) and the ends of the box cost $ 20 per square yard (A_2). The cargo box will have no salvage value and each round trip of the box on the ferry will cost 10 cents (A_3).

a) What is the minimum total cost of transporting the 400 cubic yards of gravel?
b) What are the dimensions of the cargo box?
c) What is the number of ferry trips to transport the 400 cubic yards of gravel?"

Figure 6.1 illustrates the parameters of the open cargo shipping box.

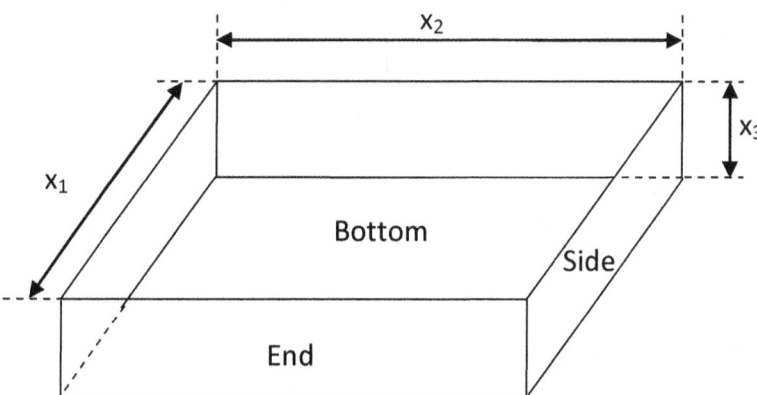

Figure 6.1: Open cargo shipping box. x_1 = Length of the Box, x_2 = Width of the Box, x_3 = Height of the Box.

6. THE OPEN CARGO SHIPPING BOX CASE STUDY

The first issue is to determine the various cost components to make the objective function. The ferry transportation cost can be determined by:

$$T1 = V * A_3/(x_1 * x_2 * x_3) = 400 * 0.10/(x_1 * x_2 * x_3) = 40/(x_1 * x_2 * x_3) \,. \tag{6.1}$$

The cost for the ends of the box (2 ends) is determined by:

$$T2 = 2 * (x_2 * x_3) * A_2 = 2 * (x_2 * x_3) * 20 = 40 * (x_2 * x_3) \,. \tag{6.2}$$

The cost for the sides of the box (2 sides) is determined by:

$$T3 = 2 * (x_1 * x_3) * A_1 = 2 * (x_1 * x_3) * 10 = 20 * (x_1 * x_3) \,. \tag{6.3}$$

The cost for the bottom of the box is determined by:

$$T4 = (x_1 * x_2) * A_1 = (x_1 * x_2) * 10 = 10 * (x_1 * x_2) \,. \tag{6.4}$$

The objective function (Y) is the sum of the four components and is:

$$Y = T1 + T2 + T3 + T4 \tag{6.5}$$
$$Y = 40/(x_1 * x_2 * x_3) + 40 * (x_2 * x_3) + 20 * (x_1 * x_3) + 10 * (x_1 * x_2) \,. \tag{6.6}$$

The primal objective function can be written in terms of generic constants for the cost variables to obtain a generalized solution.

$$Y = C_1/(x_1 * x_2 * x_3) + C_2 * (x_2 * x_3) + C_3 * (x_1 * x_3) + C_4 * (x_1 * x_2) \,, \tag{6.7}$$

where $C_1 = 40$, $C_2 = 40$, $C_3 = 20$ and $C_4 = 10$.

From the coefficients and signs, the signum values for the dual are:

$$\sigma_{01} = 1$$
$$\sigma_{02} = 1$$
$$\sigma_{03} = 1$$
$$\sigma_{04} = 1 \,.$$

The dual formulation is:

Objective Function	$\omega_{01} + \omega_{02} + \omega_{03} + \omega_{04} = 1$	(6.8)
x_1 terms	$-\omega_{01} \qquad + \omega_{03} + \omega_{04} = 0$	(6.9)
x_2 terms	$-\omega_{01} + \omega_{02} \qquad + \omega_{04} = 0$	(6.10)
x_3 terms	$-\omega_{01} + \omega_{02} + \omega_{03} \qquad = 0 \,.$	(6.11)

Using these equations, the values of the dual variables are found to be:

$$\omega_{01} = 2/5$$
$$\omega_{02} = 1/5$$
$$\omega_{03} = 1/5$$
$$\omega_{04} = 1/5$$

6.1. PROBLEM STATEMENT AND GENERAL SOLUTION

and by definition

$$\omega_{00} = 1.$$

Thus, the dual variables indicate that the first term of the primal expression is twice as important as the other three terms. The degrees of difficulty are equal to:

$$D = T - (N+1) = 4 - (3+1) = 0. \tag{6.12}$$

The objective function can be found using the dual expression:

$$Y = d(\omega) = \sigma \left[\prod_{m=0}^{M} \prod_{t=1}^{Tm} (C_{mt}\omega_{mo}/\omega_{mt})^{\sigma_{mt}\omega_{mt}} \right]^{\sigma} \tag{6.13}$$

$$= 1[[\{(C_1 * 1/(2/5))\}^{(1*2/5)}][\{C_2 * 1/(1/5)\}^{(1*1/5)}][\{C_3 * 1/(1/5)\}^{(1*1/5)}][\{C_4 * 1/(1/5)\}^{(1*1/5)}]]^1$$
$$= 100^{2/5} \quad * \quad 200^{1/5} \quad * \quad 100^{1/5} \quad * \quad 50^{1/5}$$
$$= 100^{2/5} \quad * \quad 1000000^{1/5}$$
$$= 100^{2/5} \quad * \quad 100^{3/5}$$
$$= \$100.$$

Thus, the minimum cost for transporting the 400 cubic yards of gravel across the river is $ 100. The values for the primal variables can be determined from the relationships between the primal and dual as:

$$C_1/(x_1 * x_2 * x_3) = \omega_{01} Y = (2/5)Y \tag{6.14}$$
$$C_2 * x_2 * x_3 = \omega_{02} Y = (1/5)Y \tag{6.15}$$
$$C_3 * x_1 * x_3 = \omega_{03} Y = (1/5)Y \tag{6.16}$$
$$C_4 * x_1 * x_2 = \omega_{04} Y = (1/5)Y. \tag{6.17}$$

If one combines Equations (6.15) and (6.16) one can obtain the relationship:

$$x_2 = x_1 * (C_3/C_2). \tag{6.18}$$

If one combines Equations (6.16), (6.17) and (6.18), one can obtain the relationship:

$$x_3 = x_2 * (C_4/C_3) = x_1 * (C_3/C_2) * (C_4/C_3) = x_1 * (C_4/C_2). \tag{6.19}$$

If one combines Equations (6.14) and (6.15) one can obtain the relationship:

$$x_1 * x_2^2 * x_3^2 = (1/2) * (C_1/C_2). \tag{6.20}$$

Using the values for x_2 and x_3 in Equation (6.20), one can obtain:

$$x_1 = [(1/2) * (C_1 C_2^3/(C_3^2 C_4^2))]^{1/5}. \tag{6.21}$$

Similarly, one can solve for x_2 and x_3 and the equations would be:

$$x_2 = [(1/2) * (C_1 C_3^3/(C_2^2 C_4^2))]^{1/5} \tag{6.22}$$

and

$$x_3 = [(1/2) * (C_1 C_4^3/(C_2^2 C_3^2))]^{1/5} . \tag{6.23}$$

Now using the values of $C_1 = 40$, $C_2 = 40$, $C_3 = 20$ and $C_4 = 10$, the values of x_1, x_2, and x_3 can be determined using Equations (6.21), (6.22), and (6.23) as:

$$x_1 = [(1/2) * (40 * 40^3/(20^2 10^2))]^{1/5} = [32]^{1/5} = 2 \text{ yards}$$
$$x_2 = [(1/2) * (40 * 20^3/(40^2 10^2))]^{1/5} = [1]^{1/5} = 1 \text{ yard}$$
$$x_3 = [(1/2) * (40 * 10^3/(40^2 20^2))]^{1/5} = [0.03125]^{1/5} = 0.5 \text{ yard} .$$

Thus, the box is 2 yards in length, 1 yard in width, and 0.5 yard in height. The total box volume is the product of the three dimensions, which is 1 cubic yard.

The number of trips the ferry must make is 400 cubic yards/1 cubic yard/trip = 400 trips.

If one uses the primal variables in the primal equation, the values are:

$$Y = 40/(x_1 * x_2 * x_3) + 40 * (x_2 * x_3) + 20 * (x_1 * x_3) + 10 * (x_1 * x_2) \tag{6.6}$$
$$Y = 40/(2 * 1 * 1/2) + 40 * (1 * 1/2) + 20 * (2 * 1/2) + 10 * (2 * 1) \tag{6.24}$$
$$Y = 40 + 20 + 20 + 20$$
$$= \$100$$

Note that the primal and dual give the same result for the objective function. Note that the components of the primal solution (40, 20, 20, 20) are in the same ratio as the dual variables (2/5, 1/5, 1/5, 1/5). This ratio will remain constant even as the values of the constants change and this is important in the ability to determine which of the terms are dominant in the total cost. Thus, the transportation cost is twice the cost of the box bottom and the box bottom is the same as the cost of the box sides and the same as the cost of the box ends. This indicates the optimal design relationships between the costs of the various box components and the transportation cost associated with the design.

6.2 EVALUATIVE QUESTIONS

1. The ferry cost for a round trip is increased from $ 0.10 to $ 3.20. What is the new total cost, the new box dimensions, and the number of ferry trips required to transport the 400 cubic yards of gravel?

2. The ferry cost for a round trip is increased from $ 3.20 to $ 213.06. What is the new total cost, the new box dimensions, and the number of ferry trips required to transport the 400 cubic yards of gravel.

3. A cover must be added to the box and it is made having the same costs as the box bottom. Determine the new total cost, the new box dimensions, and the number of ferry trips required to transport the 400 cubic yards of gravel.

REFERENCES

[1] R.J. Duffin, E.L. Peterson and C. Zener, *Geometric Programming*, John Wiley and Sons, New York, 1967. 21

CHAPTER 7
Metal Casting Cylindrical Riser Case Study

7.1 INTRODUCTION

The riser design problem in metal casting is always a concern for foundry engineers. The riser (also called feeders in many parts of the world) is an amount of additional metal added to a metal casting to move the thermal center of the casting and riser into the riser so there will be no solidification shrinkage in the casting. The risers are typically shaped as cylinders as other shapes are difficult for the molding process and this shape has been successfully used for decades. The riser also has other design conditions such as to supply sufficient feed metal, but thermal design issues are typically the primary concern. There are several papers on riser design using geometric programming for side riser, top riser, insulated riser and many other riser design types.

For a riser to be effective, the riser must solidify after the casting in order to provide liquid feed metal to the casting. The object is to have a riser of minimum volume to improve the yield of the casting process which improves the economics of the process. The case study considered is for a cylindrical side riser which consists of a cylinder of height H and diameter D. Figure 7.1 indicates the relationship between the casting and the side riser and the parameters of the riser.

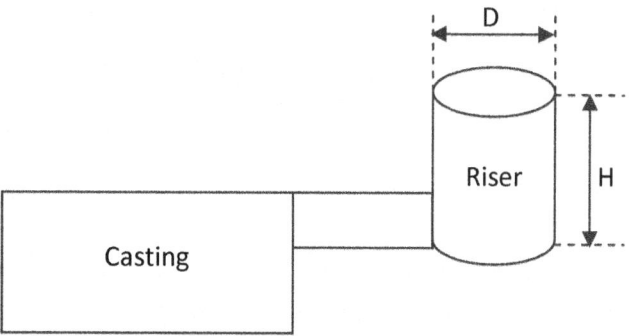

Figure 7.1: The cylindrical riser.

The theoretical basis for riser design is Chvorinov's Rule, which is:

$$t = K(V/SA)^2, \tag{7.1}$$

where

t = solidification time (minutes or seconds)
K = solidification constant for molding material (minutes/in^2 or seconds/cm^2)
V = riser volume (in^3 or cm^3)
SA = cooling surface area of the riser (in^3 or cm^3).

The objective is to design the smallest riser such that

$$t_R \geq t_C \tag{7.2}$$

where

t_R = solidification time of the riser
t_C = solidification time of the casting.

This constraint (Equation 7.2) can be written as:

$$K_R(V_R/SA_R) \geq K_C(V_C/SA_C). \tag{7.3}$$

The riser and the casting are assumed to be molded in the same material so the K_R and K_C are equal and thus the equation can be written as:

$$(V_R/SA_R) \geq (V_C/SA_C). \tag{7.4}$$

The casting has a specified volume and surface area, the right-hand side of the equation can be expressed as a constant $Y = (V_C/SA_C)$, which is called the casting modulus (M_c), and Equation (7.4) becomes

$$(V_R/SA_R) \geq Y. \tag{7.5}$$

The volume and surface of the cylindrical riser can be written as:

$$V_R = \pi D^2 H/4 \tag{7.6}$$
$$SA_R = \pi DH + 2\pi D^2/4. \tag{7.7}$$

The surface area expression neglects the connection area between the casting and the riser as the effect is small. Thus, Equation (7.5) can be rewritten as:

$$(\pi D^2 H/4)/(\pi DH + 2\pi D^2/4) = (DH)/(4H + 2D) \geq Y. \tag{7.8}$$

The constraint must be rewritten in the less than equal form with the right hand side being less than or equal to one which becomes

$$4YD^{-1} + 2YH^{-1} \leq 1. \tag{7.9}$$

7.2 PROBLEM FORMULATION AND GENERAL SOLUTION

The primal form of the side cylindrical riser design problem can be stated as:

$$\text{Minimize:} \quad V = \pi D^2 H/4 \tag{7.10}$$
$$\text{Subject to:} \quad 4YD^{-1} + 2YH^{-1} \leq 1. \tag{7.11}$$

From the coefficients and signs, the signum values for the dual are:

$$\sigma_{01} = 1$$
$$\sigma_{11} = 1$$
$$\sigma_{12} = 1$$
$$\sigma_1 = 1.$$

The dual problem formulation is:

Objective Function	ω_{01}	$= 1$	(7.12)
D terms	$2\omega_{01} - \omega_{11}$	$= 0$	(7.13)
H terms	$\omega_{01} \quad - \omega_{12}$	$= 0.$	(7.14)

Using Equations (7.12) to (7.14), the values of the dual variables were found to be:

$$\omega_{01} = 1$$
$$\omega_{11} = 2$$
$$\omega_{12} = 1.$$

The degrees of difficulty (D) are equal to:

$$D = T - (N+1) = 3 - (2+1) = 0. \tag{7.15}$$

Using the linearity inequality equation,

$$\omega_{10} = \omega_{mt} = \sigma_m \sum \sigma_{mt}\omega_{mt} = (1)*(1*2 + 1*1)$$
$$= 3 > 0 \text{ where } m = 1 \text{ and } t = 1. \tag{7.16}$$

The objective function can be found using the dual expression:

$$V = d(\omega) = \sigma \left[\prod_{m=0}^{M} \prod_{t=1}^{T_m} (C_{mt}\omega_{mo}/\omega_{mt})^{\sigma_{mt}\omega_{mt}} \right]^{\sigma} \tag{7.17}$$
$$= 1[[\{(\pi/4 * 1/1)\}^{(1*1)}][\{(4Y * 3/2)\}^{(1*2)}][\{(2Y * 3/1)\}^{(1*1)}]]^1$$
$$= (\pi/4) * (6Y)^2 * (6Y)$$
$$= (\pi/4) * (6Y)^3. \tag{7.18}$$

The values for the primal variables can be determined from the relationships between the primal and dual as:

$$4YD^{-1} = \omega_{11}/\omega_{10} = 2/3 \qquad (7.19)$$
$$\text{and} \qquad 2YH^{-1} = \omega_{11}/\omega_{10} = 1/3. \qquad (7.20)$$

The equations for H and D can be determined as:

$$D = 6Y \qquad (7.21)$$
$$\text{and} \qquad H = 6Y. \qquad (7.22)$$

Using (7.21) and (7.22) in Equation (7.10) for the primal, one obtains:

$$V = (\pi/4) * (6Y)^3. \qquad (7.23)$$

Equations (7.18) and (7.23) are identical, which is what should happen, as the primal and dual objective functions must be identical. The values for the riser diameter and the riser height are both six times the casting modulus. These relationships hold for the side cylindrical riser design with negligible effects for the connecting area. This also indicates that the riser height and riser diameter are equal for the side riser. Designs for other riser shapes and with insulating materials using geometric programming are given in the references.

7.3 EXAMPLE

A rectangular plate casting with dimensions $L = W = 10$ cm and $H = 4$ cm is to be produced and a cylindrical side riser is to be used. The optimal dimensions for the side riser can be obtained from the casting modulus Y and Equations (7.21) and (7.22). The casting modulus is obtained by:

$$Y = (V_C/SA_C)$$
$$= (10 \text{ cm} \times 10 \text{ cm} \times 4 \text{ cm})/[2(10 \text{ cm} \times 10 \text{ cm}) + 2(10 \text{ cm} \times 4 \text{ cm}) + 2(10 \text{ cm} \times 4 \text{ cm})]$$
$$= 400 \text{ cm}^3/360 \text{ cm}^2 = 1.111 \text{ cm}.$$

Thus,

$$H = 6Y = 6 \times 1.111 \text{ cm} = 6.67 \text{ cm}$$
$$D = 6Y = 6 \times 1.111 \text{ cm} = 6.67 \text{ cm}.$$

The volume of the riser can be obtained from Equation (7.23) as:

$$V = (\pi/4)(6Y)^3 = 233 \text{ cm}^3.$$

Thus, once the modulus of the casting is determined, the riser height, diameter, and volume can be determined using Equations (7.21)-(7.23).

7.4 EVALUATIVE QUESTIONS

1. A side riser is to be designed for a metal casting which has a surface area of 340 cm² and a volume of 400 cm³. The hot metal cost is 100 Rupees per kg and the metal density is 3.0 gm/cm³.

    a) What are the dimensions in centimeters for the side riser (H and D)?

    b) What is the volume of the side riser (cm³)?

    c) What is the metal cost of the side riser (Rupees)?

    d) What is the metal cost of the casting (Rupees)?

2. Instead of a side riser a top riser is to be used; that is the riser is placed on the top surface of the casting. The cooling surface area for the top riser is:

$$SA_R = \pi DH + \pi D^2/4 .$$

Show that for the top riser that $D = 6Y$ and $H = 3Y$.

REFERENCES

[1] Creese, R. C., "Optimal Riser Design by Geometric Programming," *AFS Cast Metals Research Journal,* Vol. 7, 1971, pp. 118–121.

[2] Creese, R.C., "Dimensioning of Risers for Long Freezing Range Alloys by Geometric Programming," *AFS Cast Metals Research Journal,* Vol. 7, 1971, pp. 182–184.

[3] Creese, R.C., "Generalized Riser Design by Geometric Programming," *AFS Transactions,* Vol. 87, 1979, pp. 661–664.

[4] Creese, R.C., "An Evaluation of Cylindrical Riser Designs with Insulating Materials," *AFS Transactions,* Vol. 87, 1979, pp. 665–669.

[5] Creese, R.C., "Cylindrical Top Riser Design Relationships for Evaluating Insulating Materials," *AFS Transactions*, Vol. 89, 1981, pp. 345–348.

CHAPTER 8

Inventory Model Case Study

8.1 PROBLEM STATEMENT AND GENERAL SOLUTION.

The basic inventory model is to minimize the sum of the unit set-up costs and the unit inventory holding costs. The objective is to determine the optimal production quantity which will minimize the total costs. The problem has been solved using the method of Lagrange Multipliers, but it can also easily be solved using geometric programming which permits a general solution for the production quantity in terms of the constant parameters.

The assumptions for the model are:

1. Replenishment of the order is instantaneous

2. No shortage is permitted

3. The order quantity is a batch

The model can be formulated in terms of annual costs as:

$$\text{Total Cost} = \text{Total Unit Costs} + \text{Annual Inventory Carrying Cost} + \text{Annual Set-up Cost} \tag{8.1}$$
$$TC = DC_u + CA + SD/Q . \tag{8.2}$$

where
TC = Total Annual Cost ($)
D = Annual Demand (pieces/year)
C_u = Item Unit Cost ($/piece)
C = Inventory Carrying Cost ($/piece-yr)
A = Average Inventory (Pieces)
S = Set-up Cost ($/set-up)
Q = Order Quantity (Pieces/order) .

The average inventory for this model is given by:

$$A = Q/2 . \tag{8.3}$$

Thus, the model can be formulated as:

$$TC = DC_u + CQ/2 + SD/Q \tag{8.4}$$

8. INVENTORY MODEL CASE STUDY

The model can be formulated in general terms as:

$$\text{Minimize} \quad \text{Cost}(Y) = C_{00} + C_{01}Q + C_{02}/Q \tag{8.5}$$

where

$$C_{00} = DC_u$$
$$C_{01} = C/2$$
$$C_{03} = SD.$$

Since C_{00} is a constant, the objective function can be rewritten for the variable cost as:

$$\text{Minimize} \quad \text{Variable Cost}(Y) = Y_{vc} = C_{01}Q + C_{02}/Q. \tag{8.6}$$

From the coefficients and signs, the signum values for the dual are:

$$\sigma_{01} = 1$$
$$\sigma_{02} = 1.$$

Thus, the dual formulation would be:

Objective Function (using Equation 8.6) $\quad \omega_{01} + \omega_{02} = 1 \quad$ (8.7)
Q terms (using Equation 8.6) $\quad \omega_{01} - \omega_{02} = 0 \quad$ (8.8)

The degrees of difficulty are equal to:

$$D = T - (N + 1) = 2 - (1 + 1) = 0.$$

Thus, one has the same number of variables as equations, so this can be solved by simultaneous equations as these are linear equations.

Using Equations (8.7) and (8.8), the values for the dual variables are found to be:

$$\omega_{01} = 1/2$$
$$\omega_{02} = 1/2$$

and by definition

$$\omega_{00} = 1.$$

The objective function can be found using the dual expression:

$$d(\omega) = \sigma \left[\prod_{m=0}^{M} \prod_{t=1}^{T_m} (C_{mt}\omega_{mo}/\omega_{mt})^{\sigma_{mt}\omega_{mt}} \right]^{\sigma} \tag{8.9}$$

$$d(\omega) = 1[\{(C_{01} * 1)/(1/2)\}^{(1)*(1/2)} * \{(C_{02} * 1)/(1/2)\}^{(1)*(1/2)}]^1$$
$$= 1[\{(2C_{01})^{1/2}\} * \{(2C_{02})^{1/2}\}]$$
$$= 2C_{01}^{1/2}C_{02}^{1/2}$$
$$= 2(C/2 * SD)^{1/2}$$

or
$$Y_{vc} = (2CSD)^{1/2} . \tag{8.10}$$

And
$$TC = DC_u + Y_{vc}$$

or
$$TC = DC_u + (2CSD)^{1/2} . \tag{8.11}$$

The solution has been determined without finding the value for Q. Also note that the dual expression and total cost expressions are expressed in constants and thus the answer can be found without having to resolve the entire problem as one only needs to use the new constant values.

8.2 EXAMPLE

The values for the parameters for the example problem are:

D	= Annual Demand (pieces/year)	= 100,000/yr
C_u	= Item Unit Cost ($/piece)	= $ 1.5/piece
C	= Inventory Carrying Cost ($/piece-yr)	= $0.20/piece-year
A	= Average Inventory (Pieces)	= $Q/2$
S	= Set-up Cost ($/set-up)	= $ 400/set-up
Q	= Order Quantity (Pieces/order)	= Q

Note the total cost can be found using Equation (8.11) as:
$$TC = 100,000 * 1.5 + (2 * 0.20 * 400 * 100,000)^{1/2} \tag{8.12}$$
$$= 150,000 + 4,000$$
$$= \$154,000 . \tag{8.13}$$

The value of Q can be determined from the primal-dual relationships which are:
$$(C/2)Q = \omega_{01} Y_{vc} = (1/2)(2CSD)^{1/2}$$

or
$$Q = (2SD/C)^{1/2} . \tag{8.14}$$

And for the example problem
$$Q = (2 * 400 * 100,000/0.20)^{1/2}$$
$$Q = 20,000 \text{ units} . \tag{8.15}$$

The primal problem can now be evaluated using Equation (8.4) as:
$$TC = DC_u + CQ/2 + SD/Q \tag{8.4}$$
$$TC = 100,000 * 1.5 + 0.20 * 20,000/2 + 400 * 100,000/20,000$$
$$= 150,000 + 2,000 + 2,000$$
$$= \$154,000 \tag{8.16}$$

The number of set-ups per year would be 100,000/20,000 or 5 set-ups per year.

Thus, the primal and dual give identical values for the solution of the problem as indicated by Equations (8.13) and (8.16). The dual variables are both equal to $1/2$ which implies that the total annual inventory carrying cost and the total annual set-up costs are equal as illustrated by the example.

8.3   EVALUATIVE QUESTIONS

1. The following data was collected on a new pump.

   D = Annual Demand (pieces/year) = 200/yr
   C_u = Item Unit Cost ($/piece) = $ 300/piece
   C = Inventory Carrying Cost ($/piece-yr) = $ 20/piece-year
   S = Set-up Cost ($/set-up) = $ 500/set-up

   a. Determine the total cost for the 200 pumps during the year.

   b. Determine the average total cost per pump.

   c. What is the number of set-ups per year?

   d. What is the total inventory carrying cost for the year?

2. The demand for the pumps increased dramatically to 3,000 because of the oil spill in the Gulf.

   D = Annual Demand (pieces/year) = 3,000/yr
   C_u = Item Unit Cost ($/piece) = $ 300/piece
   C = Inventory Carrying Cost ($/piece-yr) = $ 20/piece-year
   S = Set-up Cost ($/set-up) = $ 500/set-up

   a. Determine the total cost for the 3,000 pumps during the year.

   b. Determine the average total cost per pump.

   c. What is the number of set-ups per year?

   d. What is the total inventory carrying cost for the year?

3. Resolve Problem 2 if the order must be completed in one year, so recalculate the answers be since the number of set-ups was not an integer?

REFERENCES

[1] H. Jung and C.M. Klein, "Optimal Inventory Policies Under Decreasing Cost Functions via Geometric Programming", *European Journal of Operational Research*, 132, 2001, pp. 628–642.

CHAPTER 9

Process Furnace Design Case Study

9.1 PROBLEM STATEMENT AND SOLUTION

An economic process model was developed [1, 2] for an industrial metallurgical application. The annual cost for a furnace operation in which the slag-metal reaction is a critical factor of the process was considered and a modified version of the problem is presented. The objective was to minimize the annual cost and the primal equation representing the model was:

$$Y = C_1/(L^2 * D * T^2) + C_2 * L * D + C_3 * L * D * T^4 . \tag{9.1}$$

The model was subject to the constraint that:

$$D \leq L .$$

The constraint must be set in geometric programming for which would be:

$$(D/L) \leq 1 \tag{9.2}$$

where

$$D = \text{Depth of the furnace (ft)}$$
$$L = \text{Characteristic Length of the furnace (ft)}$$
$$T = \text{Furnace Temperature (K)} .$$

For the specific example problem, the values of the constants were:

$$C_1 = 10^{13}(\$ - \text{ft}^2 - \text{K}^2)$$
$$C_2 = 100(\$/\text{ft}^2)$$
$$C_3 = 5 * 10^{-11}(\text{ft}^{-2} - \text{K}^{-4}) .$$

From the coefficients and signs, the signum values for the dual are:

$$\sigma_{01} = 1$$
$$\sigma_{02} = 1$$
$$\sigma_{03} = 1$$
$$\sigma_{11} = 1$$
$$\sigma_1 \;= 1 .$$

9. PROCESS FURNACE DESIGN CASE STUDY

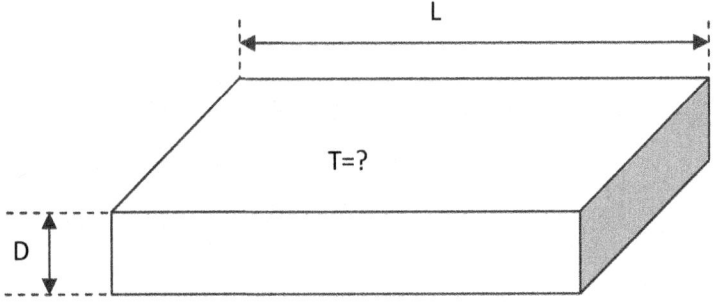

Figure 9.1: Process furnace.

The dual problem formulation is:

Objective Function	$\omega_{01} + \omega_{02} + \omega_{03} = 1$	(9.3)
L terms	$-2\omega_{01} + \omega_{02} + \omega_{03} - \omega_{11} = 0$	(9.4)
D terms	$-\omega_{01} + \omega_{02} + \omega_{03} + \omega_{11} = 0$	(9.5)
T terms	$-2\omega_{01} \phantom{+ \omega_{02}} + 4\omega_{03} \phantom{+ \omega_{11}} = 0.$	(9.6)

Using Equations (9.3) to (9.6), the values of the dual variables were found to be:

$$\omega_{01} = 0.4$$
$$\omega_{02} = 0.4$$
$$\omega_{03} = 0.2$$
$$\omega_{11} = -0.2.$$

The dual variables cannot be negative and the negative value implies that the constraint is not binding, that is it is a loose constraint. Thus, the problem must be reformulated without the constraint and the dual variable is forced to zero, that is, $\omega_{11} = 0$ and the equations resolved. This means that the constraint $D \leq L$ will be loose, that is D will be less than L in the solution. The new dual becomes:

Objective Function	$\omega_{01} + \omega_{02} + \omega_{03} = 1$	(9.7)
L terms	$-2\omega_{01} + \omega_{02} + \omega_{03} = 0$	(9.8)
D terms	$-\omega_{01} + \omega_{02} + \omega_{03} = 0$	(9.9)
T terms	$-2\omega_{01} \phantom{+ \omega_{02}} + 4\omega_{03} = 0.$	(9.10)

Now the problem is that it has 4 equations to solve for three variables. If one examines Equations (9.8) and (9.9), one observes that Equation (9.8) is dominant over Equation (9.9) and thus Equation (9.9) will be removed from the dual formulation. The new dual formulation is:

9.1. PROBLEM STATEMENT AND SOLUTION

$$\text{Objective Function} \quad \omega_{01} + \omega_{02} + \omega_{03} = 1 \quad (9.11)$$
$$L \text{ terms} \quad -2\omega_{01} + \omega_{02} + \omega_{03} = 0 \quad (9.12)$$
$$T \text{ terms} \quad -2\omega_{01} \phantom{+ \omega_{02}} + 4\omega_{03} = 0. \quad (9.13)$$

The new solution for the dual becomes:

$$\omega_{01} = 1/3$$
$$\omega_{02} = 1/2$$
$$\omega_{03} = 1/6$$

and by definition

$$\omega_{00} = 1.0.$$

The dual variables indicate that the second term is the most important, followed by the first term and then the third term. The degrees of difficulty are now equal to:

$$D = T - (N + 1) = 3 - (2 + 1) = 0.$$

The objective function can be found using the dual expression:

$$Y = d(\omega) = \sigma \left[\prod_{m=0}^{M} \prod_{t=1}^{T_m} (C_{mt}\omega_{mo}/\omega_{mt})^{\sigma_{mt}\omega_{mt}} \right]^{\sigma} \quad (9.14)$$

$$= 1[[\{(C_1 * 1/(1/3)1)\}^{(1/3*1)}][\{(C_2 * 1/(1/2)\}^{(1/2*1)}][\{(C_3/(1/6))\}^{(1/6*1)}]]^1$$

$$= 1[[\{(1 * 10^{13} * 1/(1/3)1)\}^{(1/3*1)}][\{(100 * 1/(1/2)\}^{(1/2*1)}][\{(5 * 10^{-11}/(1/6))\}^{(1/6*1)}]]^1$$

$$= \$11,370.$$

This can be expressed in a general form in terms of the constants as:

$$Y = (3C_1)^{1/3}(2C_2)^{1/2}(6C_3)^{1/6}. \quad (9.15)$$

The values for the primal variables can be determined from the relationships between the primal and dual which are:

$$C_1 * L^{-2} * D^{-1} * T^{-2} = \omega_{01}Y \quad (9.16)$$
$$C_2 * L * D = \omega_{02}Y \quad (9.17)$$
$$\text{and} \quad C_3 * L * D * T^4 = \omega_{03}Y. \quad (9.18)$$

9. PROCESS FURNACE DESIGN CASE STUDY

The fully general expressions are somewhat difficult, but the variables can be expressed in terms of the constants and objective function as:

$$T = [(\omega_{03}/\omega_{02}) * (C_2/C_3)]^{1/4} \tag{9.19}$$
$$L = [(C_1 * (C_2 * C_3)^{(1/2)})]/[\omega_{01} * (\omega_{02} * \omega_{03})^{(1/2)} * Y^2] \tag{9.20}$$
$$D = [(\omega_{01} * \omega_{02}^{(3/2)} * \omega_{03}^{(1/2)} Y^3)/(C_1 * C_2^{(3/2)} * C_3^{(1/2)}]. \tag{9.21}$$

The expressions developed for the variables in terms of the constants in a reduced form were:

$$T = (C_2/3C_3)^{1/4} \tag{9.22}$$
$$L = (3C_1)^{1/3}(2C_2)^{-1/2}(6C_3)^{1/6} \tag{9.23}$$
$$D = 1. \tag{9.24}$$

Using the values of $Y = 11370$, $C_1 = 10^{13}$, $C_2 = 100$, $C_3 = 5 * 10^{-11}$, $\omega_{01} = 1/3$, $\omega_{02} = 1/2$ and $\omega_{03} = 1/6$, the values for the variables are:

$$T = 903 \text{ K}$$
$$L = 56.85 \text{ ft}$$

and

$$D = 1.00 \text{ ft.}$$

Using the values of the variables in the primal equation, the objective function is:

$$Y = C_1/(L^2 * D * T^2) + C_2 * L * D + C_3 * L * D * T^4$$
$$= 10^{13}/(56.85^2 * 1 * 903^2) + 100 * 56.85 * 1 + 5 * 10^{-11} * 56.85 * 1 * (903^4)$$
$$= 3,795 + 5,685 + 1,890$$
$$= \$11,370.$$

The values of the objective function for the primal and dual are identical, which implies that the values for the primal variables have been correctly obtained. The costs terms are in the same ratio as the dual variables; the third term is the smallest, the first term is twice the third term and the second term is three times the third term.

This problem was presented to indicate the difficulties in that when the constraint is loose, the problem must be restated with the loose constraint removed and the new dual variables are resolved. The constraint is loose as $D = 1$ ft is much lower than $L = 56.85$ ft. The other item of interest was that equations dominated by other equations can prevent a solution and must be removed. The removal of the dominated equation was necessary to obtain a solution and may be the cause of D being unity.

9.2 EVALUATIVE QUESTIONS

1. The problem constraint was given as $D \leq L$, but the designer decided that was incorrect and reversed the constraint to $L \leq D$. Resolve the problem and determine the dual and primal variables as well as the objective function.

2. Resolve the problem making the initial assumption that $L = D$ and reformulate the primal and dual problems and find the variables and objective function.

REFERENCES

[1] http://www.mpri.lsu/edu/textbook/Chapter3-b.htm (visited May 2009). 37

[2] Ray, W.H. and J Szekely, *Process Optimization with Applications in Metallurgy and Chemical Engineering*, John Wiley and Sons, Inc., New York (1973). 37

CHAPTER 10

Gas Transmission Pipeline Case Study

10.1 PROBLEM STATEMENT AND SOLUTION

The energy crisis is with us today, and one of the problems is in the transmission of energy. A gas transmission model was developed [1, 2] to minimize the total transmission cost of gas in a new gas transmission pipeline. The problem is more difficult than the previous case studies as several of the exponents are not integers. The primal expression for the cost developed was:

$$C = C_1 * L^{1/2} * V/(F^{0.387} * D^{2/3}) + C_2 * D * V + C_3/(L * F) + C_4 * F/L . \qquad (10.1)$$

Subject to:

$$(V/L) \geq F .$$

The constraint must be restated in the geometric form as:

$$-(V/(LF)) \leq -1 , \qquad (10.2)$$

where
  L = Pipe length between compressors (feet)
  D = Diameter of Pipe (in)
  V = Volume Flow Rate (ft^3/sec)
  F = Compressor Pressure Ratio Factor.

Figure 10.1 is a sketch of the problem indicating the variables and is not drawn to scale.

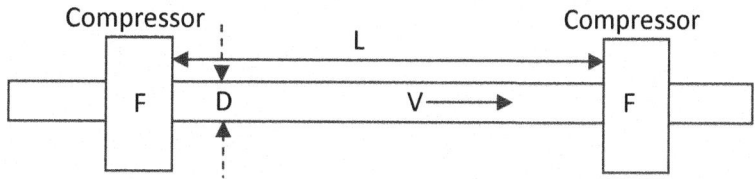

Figure 10.1: Gas transmission pipeline.

10. GAS TRANSMISSION PIPELINE CASE STUDY

For the specific problem, the values of the constants were:

$$C_1 = 4.55 * 10^5$$
$$C_2 = 3.69 * 10^4$$
$$C_3 = 6.57 * 10^5$$
$$C_4 = 7.72 * 10^5.$$

From the coefficients and signs, the signum values for the dual are:

$$\sigma_{01} = 1$$
$$\sigma_{02} = 1$$
$$\sigma_{03} = 1$$
$$\sigma_{04} = 1$$
$$\sigma_{11} = -1$$
$$\sigma_1 = -1.$$

The dual problem formulation is:

Objective Function	$\omega_{01} + \omega_{02} + \omega_{03} + \omega_{04} = 1$		(10.3)
L terms	$0.5\omega_{01} \quad - \omega_{03} - \omega_{04} + \omega_{11} = 0$		(10.4)
F terms	$-0.387\omega_{01} \quad - \omega_{03} + \omega_{04} + \omega_{11} = 0$		(10.5)
V terms	$\omega_{01} + \omega_{02} \quad - \omega_{11} = 0$		(10.6)
D terms	$-0.667\omega_{01} + \omega_{02} \quad = 0.$		(10.7)

Using Equations (10.3) to (10.7), the values of the dual variables were found to be:

$$\omega_{01} = 0.26087$$
$$\omega_{02} = 0.17391$$
$$\omega_{03} = 0.44952$$
$$\omega_{04} = 0.11570$$
$$\omega_{11} - 0.43478$$

and by definition
$$\omega_{00} = 1,$$
and

$$\omega_{10} = \omega_{mt} = \sigma_m \sum \sigma_{mt}\omega_{mt} = (-1)*(-1*0.43478) = 0.43478 \text{ where } m = 1 \text{ and } t = 1.$$

10.1. PROBLEM STATEMENT AND SOLUTION

The objective function can be found using the dual expression:

$$Y = d(\omega) = \sigma \left[\prod_{m=0}^{M} \prod_{t=1}^{T_m} (C_{mt}\omega_{mo}/\omega_{mt})^{\sigma_{mt}\omega_{mt}} \right]^{\sigma} \tag{10.8}$$

$$= 1[[\{(4.55*10^5*1/0.26087)\}^{(1*0.26087)}]*[\{(3.69*10^4*1/0.17391)\}^{(1*0.17391)}]*$$
$$[\{(6.57*10^5*1/0.44952)\}^{(1*0.44952)}]*[\{(7.72*10^5*1/0.11570)\}^{(1*0.11570)}]*$$
$$[\{(1*0.43478/0.43478)\}^{(-1*0.43478)}]]^1$$
$$= \$1.3043*10^6/\text{yr}.$$

The degrees of difficulty are equal to:

$$D = T - (N+1) = 5 - (4+1) = 0. \tag{10.9}$$

The values for the primal variables can be determined from the relationships between the primal and dual which are:

$$C_1 * L^{1/2} * V/(F^{0.387} * D^{2/3}) = \omega_{01}Y \tag{10.10}$$
$$C_2 * D * V = \omega_{02}Y \tag{10.11}$$
$$C_3/(L*F) = \omega_{03}Y \tag{10.12}$$
$$C_4 * F/L = \omega_{04}Y \tag{10.13}$$
$$V/(F*L) = \omega_{11}/\omega_{10} = 1. \tag{10.14}$$

The fully general expressions are somewhat difficult, but the variables can be expressed in terms of the constants and objective function as:

$$F = [(C_3\omega_{04})/(C_4\omega_{03})]^{1/2} \tag{10.15}$$
$$V = C_3/(\omega_{03} * Y) \tag{10.16}$$
$$L = [(C_3 C_4)/(\omega_{03}\omega_{04})]/Y \tag{10.17}$$
$$D = [(\omega_{02} * \omega_{03})/(C_2 * C_3)] * Y^2. \tag{10.18}$$

Using the values of $Y = 1.3043 * 10^6$, $C_1 = 4.55 * 10^5$, $C_2 = 3.69 * 10^4$, $C_3 = 6.57 * 10^5$, $C_4 = 7.72 * 10^5$, $\omega_{01} = 0.26087$, $\omega_{02} = 0.17391$, $\omega_{03} = 0.44952$, $\omega_{04} = 0.11570$, and $\omega_{11} = 0.43478$ one obtains:

$$F = [(C_3 * \omega_{04})/(C_4 * \omega_{03})]^{1/2} = [(6.57*10^5*0.11570)/(7.72*10^5*0.44952)]^{1/2} = 0.468$$
$$V = C_3/(\omega_{03}*Y) = 6.57*10^5/(0.44952*1.3043*10^6) = 1.1205 \text{ ft}^3/\text{sec}$$
$$L = [(C_3*C_4)/(\omega_{03}*\omega_{04})]^{1/2}/Y = [(6.57*10^5*7.72*10^5)/(0.44952*0.1157)]^{1/2}/1.3043*10^6 = 2.3943 \text{ f}$$
$$D = [(\omega_{02}*\omega_{03})/(C_2*C_3)]*Y^2 = [(0.17391*0.44952)(3.69*10^4 6.57*10^5)]*(1.3043*10^6)^2 = 5.4857 \text{ in}.$$

10. GAS TRANSMISSION PIPELINE CASE STUDY

The primal expression can now be solved using the primal variables and the contribution of each of the terms can be observed.

$$\begin{aligned}
C &= C_1 * L^{1/2} * V/(F^{0.387} * D^{2/3}) + C_2 * D * V + C_3/(L*F) + C_4 * F/L \\
&= 4.55 * 10^5 * 2.3943^{1/2} * 1.1205/(0.468^{0.387} * 5.4857^{2/3}) + 3.69 * 10^4 * 5.4857 * 1.1205 + \\
&\quad 6.57 * 10^5/(2.3943 * 0.468) + 7.72 * 10^5 * 0.468/2.3943 \\
&= 3.4029 * 10^5 + 2.26814 * 10^5 + 5.8633 * 10^5 + 1.5090 * 10^5 \\
&= \$1,304,300 .
\end{aligned}$$

The third term is slightly higher than the others, but all terms are of the same magnitude. Since the constraint is binding, that is $V = L * F$, and the results indicate that holds as:

$$1.1205 = 2.3943 * 0.468 = 1.1205 .$$

The values of the dual variables were more complex for this problem than the previous problems, but the values of these dual variables still have the same relationship to the terms of the primal cost function. The first dual variable, ω_{01} was 0.26087, and the relation between the first cost term of the primal to the total cost is $3.4029 * 10^5/1.3043 * 10^6 = 0.2609$. The reader should show that the other dual variables have the same relationships between the terms of the primal cost function and the total primal cost.

10.2 EVALUATIVE QUESTIONS

1. Resolve the problem with the values of $C_1 = 6 * 10^5$, $C_2 = 5 * 10^4$, $C_3 = 7 * 10^5$, and $C_4 = 8 * 10^5$. Determine the effect upon the dual variables, the objective function, and the primal variables. Also examine the percentage of each of the primal terms in the objective function and in the original objective function.

2. The constraint is a binding constraint. If the constraint is removed, the objective function should be lower. What problem(s) occurs when the constraint is removed that causes concern?

REFERENCES

[1] http://www.mpri.lsu.edu/textbook/Chapter3-b.htm (visited May 2009). 43

[2] Sherwood, T.K., *A Course in Process Design*, MIT Press, Cambridge, Mass. (1963). 43

CHAPTER 11
Profit Maximization Case Study

11.1 PROFIT MAXIMIZATION AND GEOMETRIC PROGRAMMING

The examples thus far have been minimization problems and the maximization problems are slightly different. In minimization problems, the terms of the objective function tend to have positive signs or all positive Signum functions. The constraints may have negative signum functions in a minimization problem, but those in the objective function tend to be positive. In the profit maximization problem, the revenues have positive coefficients and the costs have negative coefficients. The solution formulation for the maximization problem is slightly different than the minimization problem and a solution to an example is presented to illustrate the problem formulation and solution procedure.

11.2 PROFIT MAXIMIZATION USING THE COBB-DOUGLAS PRODUCTION FUNCTION

The problem selected was presented by Liu [1] who adapted it from Keyzer and Wesenbeeck [2]. The objective is to maximize the profit, π, using a Cobb-Douglas production function and the component costs. The problem given is

$$\text{Maximize} \quad \pi = pAx_1^{0.1} x_2^{0.3} x_3^{0.2} - C_{01}x_1 - C_{02}x_2 - C_{03}x_3 \tag{11.1}$$

where

Parameter	Term	Value for Problem
p	= market price	= 20
A	= scale of production(Cobb-Douglas function)	= 40
C_{01}	= Cost of Product 1	= 20
C_{02}	= Cost of Product 2	= 24
C_{03}	= Cost of Product 3	= 30
C_{00}	= pA	= 800

To solve the problem, one minimizes the negative of the profit function; that is the primal objective function is:

$$\text{Minimize} \quad Y = -C_{00}x_1^{0.1} x_2^{0.3} x_3^{0.2} + C_{01}x_1 + C_{02}x_2 + C_{03}x_3 \tag{11.2}$$

were $Y = -\pi$.

11. PROFIT MAXIMIZATION CASE STUDY

The dual objective function would be:

$$D(Y) = -1[(C_{00}/\omega_{01})_{01}^{-\omega}(C_{01}/\omega_{02})_{02}^{\omega}(C_{02}/\omega_{03})_{03}^{\omega}(C_{03}/\omega_{04})_{04}^{\omega}]^{-1}. \tag{11.3}$$

For maximization of profits, the signum function $\sigma_{00} = -1$, whereas in the dual objective function for the minimization of costs the signum function $\sigma_{00} = 1$, a positive value. From the coefficients and signs, the signum values for the dual are:

$$\sigma_{01} = -1$$
$$\sigma_{02} = 1$$
$$\sigma_{03} = 1$$
$$\sigma_{04} = 1.$$

The dual formulation would be:

Objective Fctn	$-\omega_{01}$	$+\omega_{02}$	$+\omega_{03}$	$+\omega_{04} = -1$(max)	(11.4)
x_1 terms	$-0.1\omega_{01}$	$+\omega_{02}$		$= 0$	(11.5)
x_2 terms	$-0.3\omega_{01}$		$+\omega_{03}$	$= 0$	(11.6)
x_3 terms	$-0.2\omega_{01}$			$+\omega_{04} = 0$	(11.7)

Solving Equations (11.4) through (11.7) for the dual variables one obtains:

$$\omega_{01} = 2.50$$
$$\omega_{02} = 0.25$$
$$\omega_{03} = 0.75$$
$$\omega_{04} = 0.50.$$

The dual variables do not sum to unity which was the case for the cost minimization problems. However, the difference between the profit dual variable and cost dual variables is equal to unity. Also, the cost terms should be in the same ratio as their corresponding dual variables. The dual objective function would result in

$$D(Y) = -1[(800/2.5)^{-2.5}(20/0.25)^{0.25}(24/0.75)^{0.75}(30/0.50)^{0.50}]^{-1} \tag{11.8}$$
$$= -5,877.12.$$

The primal variables can be found from the primal-dual relationships and the values of the dual objective function and dual variables similar to that in the cost minimization procedure.

Thus,

$$x_1 = \omega_{02}Y/C_{02} = 0.25 \times 5877.12/20 = 73.46 \tag{11.9}$$
$$x_2 = \omega_{03}Y/C_{03} = 0.75 \times 5877.12/24 = 183.66 \tag{11.10}$$
$$x_3 = \omega_{04}Y/C_{04} = 0.50 \times 5877.12/30 = 97.95. \tag{11.11}$$

If the values for x_1, x_2 and x_3 are used in the primal, one obtains:

$$\text{Maximize} \quad \pi = pAx_1^{0.1}x_2^{0.3}x_3^{0.2} - C_{01}x_1 - C_{02}x_2 - C_{03}x_3 \quad (11.1)$$
$$= 800(73.46)^{0.1}(183.66)^{0.3}(97.95)^{0.2} - 20 \times 73.46 - 24 \times 183.66 - 30 \times 97.95$$
$$= 14,692.66 - 1469.2 - 4,407.84 - 2,938.5$$
$$\pi = 5,877.12$$

Therefore, the values of the primal and dual objective functions are equal. The three cost terms of 1469.2, 4,407.84, and 2.938.5 are in the same ratio as the cost dual variables of 0.25, 0.75, and 0.50 or 1:3:2. The primal variables, x_1, x_2 and x_3, are not in the same ratio as the dual variables.

11.3 EVALUATIVE QUESTIONS

1. What is the history of the Cobb-Douglas Production Function?

2. What is the primary difference between the dual objective functions for the minimum cost and maximum profit models?

3. The profit function is:

$$\text{Maximize } \pi = pAx_1^{0.1}x_2^{0.3}x_3^{0.2} - C_{01}x_1 - C_{02}x_2 - C_{03}x_3^{0.80}$$

And the values for the primal coefficients are:

$$pA = \$500$$
$$C_{01} = \$40$$
$$C_{02} = \$30$$
$$C_{03} = \$20$$

   (a) Solve for the dual variables and the dual objective function.
   (b) Solve for the primal variables and the primal objective function.
   (c) Show that the cost terms of the primal are in the same ratio as the dual variables.

4. The following problem is from the web site http://www.mpri.lsu.edu/textbook/Chapter3-b.htm

   Maximize the function:

$$Y = 3x_1^{0.25} - 3x_1^{1.1}x_2^{0.6} - 115x_2^{-1}x_3^{-1} - 2x_3 .$$

   (a) Solve for the dual variables and the dual objective function.
   (b) Solve for the primal variables and the primal objective function.
   (c) Show that the cost terms (negative functions) of the primal are in the same ratio as the dual variables.

   (Hint – solve the dual in terms of fractions answers rather than decimal values.)

REFERENCES

[1] Shiang-Tai Liu, "A Geometric Programming Approach to Profit Maximization", *Applied Mathematics and Computation*, 182, (2006), pp 1093–1097. 47

[2] M. Keyzer and L. Wesenbeck, "Equilibrium Selection in Games; the Mollifier Method", *Journal of Mathematical Economics*, 41, (2005), pp 285–301. 47

CHAPTER 12

Material Removal/Metal Cutting Economics Case Study

12.1 INTRODUCTION

Material removal economics, also known as metal cutting economics or machining economics, is an example of a problem which has non-integer exponents, and this makes the problem challenging. This problem has been presented previously [1, 2], but this version is slightly different from, and easier than, those presented earlier. The material removal economics problem is based upon the Taylor Tool Life Equation, which was developed by Frederick W. Taylor over 100 years ago in the USA. There are several versions of the equation, and the form selected is one of the modified versions which includes cutting speed and feed rate. The equation selected was:

$$TV^{1/n}f^{1/m} = C , \qquad (12.1)$$

where

T = tool life (minutes)
V = cutting speed (ft/min or m/min)
F = feed rate (inches/rev or mm/rev)
$1/n$ = cutting speed exponent
$1/m$ = feed rate exponent
C = Taylor's Modified Tool Life Constant (ft/min or m/min).

The object is to minimize the total cost for machining, operator, tool cost and tool changing cost.

12.2 PROBLEM FORMULATION

An expression for the machining cost, operator cost, tool cost and tool changing cost was developed [1] and the resulting expression was:

$$C_u = K_{00} + K_{01}f^{-1}V^{-1} + K_{02}f^{(1/m-1)}V^{(1/n-1)} , \qquad (12.2)$$

where

C_u = total unit cost
K_{00} = $(R_o + R_m)t_l$
K_{01} = $(R_o + R_m)B$
K_{02} = $[(R_o + R_m)t_{ch} + C_t]QBC^{-1}$,

and

- R_o = operator rate ($/min), derived from Operator's Hourly Cost
- R_m = machine rate ($/min), derived from Machine Hour Cost
- t_l = machine loading & unloading time (min)
- t_{ch} = tool changing time (min)
- B = cutting path surface factor of tool (in-ft, or mm-m)
- Q = fraction of cutting path that tool is cutting material
- C_t = tool cost ($/cutting edge), derived from Tool Insert Cost
- C = Taylor's Modified Tool Life Constant (min).

The first part of the objective function expression represents the loading and unloading costs, the second part represents the cutting costs, and the third part represents the tool and tool changing costs. The loading and unloading costs are not a function of the feed and cutting speed. Since K_{00} is a constant, the primal problem can be formulated as solving for the variable cost, $C_u(\text{var})$ as:

$$C_u(\text{var}) = K_{01} f^{-1} V^{-1} + K_{02} f^{(1/m-1)} V^{1/n-1} . \tag{12.3}$$

Subject to a maximum feed constraint written as:

$$K_{11} f \leq 1 , \tag{12.4}$$

where

$$K_{11} = 1/f_{\max} .$$

From the coefficients and signs, the signum values for the dual are:

$$\sigma_{01} = 1$$
$$\sigma_{02} = 1$$
$$\sigma_{11} = 1$$
$$\sigma_1 = 1 .$$

The dual problem formulation is:

Objective Function	$\omega_{01} + \omega_{02}$	$= 1$	(12.5)
f terms	$-\omega_{01} + (1/m - 1)\omega_{02} + \omega_{11}$	$= 0$	(12.6)
V terms	$-\omega_{01} + (1/n - 1)\omega_{02}$	$= 0 .$	(12.7)

The degrees of difficulty are equal to:

$$D = T - (N+1) = 3 - (2+1) = 0 . \tag{12.8}$$

From the constraint equations, which have only one term, it is apparent that:

$$\omega_{10} = \omega_{11} . \tag{12.9}$$

Since there are zero degrees of difficulty, the dual parameters can be solve directly. Thus, if one adds Equations (12.5) and (12.7), one can solve for ω_{02} directly and obtain:

$$\omega_{02} = n . \tag{12.10}$$

Then, from Equation (12.5), one obtains;

$$\omega_{01} = 1 - n . \tag{12.11}$$

Finally, by using the values for ω_{01} and ω_{02}, one can determine ω_{11} as:

$$\omega_{11} = 1 - n/m . \tag{12.12}$$

The objective function can be evaluated using the dual expression:

$$C_u(\text{var}) = d(\omega) = \sigma \left[\prod_{m=0}^{M} \prod_{t=1}^{T_m} (C_{mt} \omega_{mo}/\omega_{mt})^{\sigma_{mt}\omega_{mt}} \right]^\sigma \tag{12.13}$$

$$C_u(\text{var}) = 1\{[(K_{01}\omega_{00}/\omega_{01})^{\omega_{01}}][(K_{02}\omega_{00}/\omega_{02})^{\omega_{02}}][(K_{11}\omega_{10}/\omega_{11})^{\omega_{11}}]\}^1 \tag{12.14}$$
$$= [(K_{01}1/(1-n))^{(1-n)}][(K_{02}1/n)^n][(K_{11}(1-n/m)/(1-n/m))^{(1-n/m)}]$$
$$= [K_{01}/(1-n)][(K_{02}/K_{01})((1-n)/n)]^n[(K_{11})^{(1-n/m)}]$$
$$= (K_{11})^{1-n/m} K_{01}^{1-n} K_{02}^n (1-n)^{n-1}/n^n . \tag{12.15}$$

The primal variables, V and f, can be evaluated from the primal-dual relationships.

$$K_{01} f^{-1} V^{-1} = \omega_{01} C_u(\text{var}) \tag{12.16}$$
$$K_{02} f^{1/m-1} V^{1/n-1} = \omega_{02} C_u(\text{var}) \tag{12.17}$$
$$K_{11} f = 1 . \tag{12.18}$$

From Equation (12.18), is seen that:

$$f = 1/K_{11} . \tag{12.19}$$

If one divides Equation (12.17) by Equation (12.16), one obtains:

$$f^{1/m} V^{1/n} = (K_{01}/K_{02})(n/(1-n)) . \tag{12.20}$$

Using Equation (12.19) in (12.20) and solving for V, one obtains

$$V = [(n/(1-n))]^n (K_{01}/K_{02})^n K_{11}^{n/m} . \tag{12.21}$$

Now, if one uses the values of f and V from Equations (12.19) and (12.21) in the primal Equation (12.3) and also using Equation (12.20), one obtains:

$$C_u(\text{var}) = K_{01} f^{-1} V^{-1} + K_{02} f^{(1/m-1)} V^{(1/n-1)} , \tag{12.3}$$
$$= f^{-1} V^{-1} [K_{01} + K_{02} f^{1/m} V^{1/n}]$$
$$= K_{11}[(n/(1-n))^{-n} (K_{01}/K_{02})^{-n} K_{11}^{-n/m} [K_{01} + K_{02}(n/(1-n))(K_{01}/K_{02})]$$
$$= K_{11}^{1-n/m} (n/(1-n))^{-n} (K_{01}/K_{02})^{-n} [K_{01} + K_{01}(n/(1-n))]$$
$$= K_{11}^{1-n/m} K_{01}^{1-n} K_{02}^n (1-n)^{n-1}/n^n . \tag{12.22}$$

54 12. MATERIAL REMOVAL/METAL CUTTING ECONOMICS CASE STUDY

The variable unit cost expressions, C_u (var), are identical for both the primal and dual formulations. The expressions for the primal variables and the variable unit cost are more complex than the expressions obtained in the previous models because of the non-integer exponents.

Elaborate research work has been done with the material removal problems, and the dissertation by Pingfang Tsai [3] has solutions for problems with an additional variable, the depth of cut, and additional constraints on horsepower of the motor driving the main spindle of the lathe and depth of cut. With the additional constraints, there is the possibility of loose constraints, and a flow chart has been developed for the different solutions depending upon which constraints are loose. Chapter 16 presents a material removal problem with one degree of difficulty and two constraints.

12.3 EVALUATIVE QUESTIONS

1. A cylindrical bar, 6 inches long and 1 inch in diameter is to be finished turned on a lathe. The maximum feed to be used to control the surface finish is 0.005 in/rev. Find the total cost to machine the part, the variable cost to machine the part, the feed rate, the cutting speed, and the tool life in minutes. Use both the primal and dual equations to determine the variable unit cost. The data are:

   R_o = 0.60 \$/min
   R_m = 0.40 \$/min
   C_t = \$ 2.00/edge
   t_l = 1.5 min
   $t_c h$ = 0.80 min
   D = 1 inch
   L = 6 inches
   $1/m$ = 1.25 (m = 0.80)
   $1/n$ = 4.00 (n = 0.25)
   C = 5.0 x 10^8 min
   Q = 1.0 (for turning).

   Using these values, one can obtain:

   K_{00} = 1.50
   K_{01} − 1.57 in-ft
   K_{02} = 8.8 x 10^{-9}
   (solution f = 0.005 in/rev, V = 459 ft/min, C_u (var) = 0.91, and T = 8.5 min).

2. A cylindrical bar, 150 mm long and 25 mm in diameter is to be finished turned on a lathe. The maximum feed to be used to control the surface finish is 0.125 mm/rev. Find the total cost to machine the part, the variable cost to machine the part, the feed rate, the cutting speed, and the tool life in minutes. The data are:

R_o = 0.60 \$/min
R_m = 0.40 \$/min
C_t = \$ 2.00/edge
t_l = 1.5 min
t_ch = 0.8 min
D = 25 mm
L = 150 mm
$1/m$ = 1.25 (m = 0.80)
$1/n$ = 4.00 (n = 0.25)
C = 2.46 x 10^8 min
Q = 1.0 (for turning).

Using these values, one can obtain:

K_{00} = 1.50
K_{01} = 11.78 mm-m
K_{02} = 1.34 x 10^{-7}

(solution f = 0.125 mm/rev, $V = 140$ m/min, C_u (var) = 0.90, and $T = 8.6$ min).

REFERENCES

[1] Robert C. Creese and Pingfang Tsai, "Generalized Solution for Constrained Metal Cutting Economics Problem," *1985 Annual International Industrial Conference Proceedings*, Institute of Industrial Engineers U.S.A, pp 113–117. 51

[2] Ermer, D.S., "Optimization of the Constrained Machining Economics Problem by Geometric Programming," *Journal of Engineering for Industry*, Transactions of the ASME, November 1971, pp 1067–1072. 51

[3] Tsai, Pingfang *An Optimization Algorithm and Economic Analysis for a Constrained Machining Model*, PhD Dissertation, West Virginia University, Morgantown, WV. 54

PART III

Geometric Programming Applications with Positive Degrees of Difficulty

CHAPTER 13

Journal Bearing Design Case Study

13.1 INTRODUCTION

An interesting problem with one degree of difficulty is a journal bearing design problem presented by Beightler, Lo, and Bylander [1]. The objective was to minimize the cost (P), and the variables were the half-length of the bearing (L) and the radius of the journal (R). The objective function and the constants in the problem are those presented in the original paper, and the derivations of the constants were not detailed. The solution is based upon deriving an additional equation whereas the original problem was solved by reducing the degree of difficulty and determining upper and lower bounds to the solution. The solution presented solves the problem, directly using the additional equation and without needing to use search techniques.

13.2 PRIMAL AND DUAL FORMULATION OF JOURNAL BEARING DESIGN

The generalized primal problem was:

$$\text{Minimize} \quad P = C_{01} R^3 L^{-2} + C_{02} R^{-1} + C_{03} R L^{-3} \qquad (13.1)$$
$$\text{Subject to:} \ C_{11} * R^{-1} * L^3 \leq 1, \qquad (13.2)$$

where
- P = Cost ($)
- R = radius of the journal (in)
- L = half-length of the bearing (in)
- C_{01} = 0.44 (for example problem)
- C_{02} = 10 (for example problem)
- C_{03} = 0.592 (for example problem)
- and C_{11} = 8.62 (for example problem).

Figure 13.1 is a sketch illustrating the variables for the problem.

From the coefficients and signs, the signum values for the dual from Equations (13.1) and (13.2) are:

13. JOURNAL BEARING DESIGN CASE STUDY

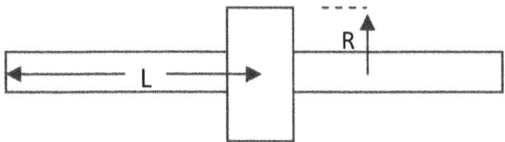

Figure 13.1: Journal bearing parameters of half-lengh and radius.

$$\sigma_{01} = 1$$
$$\sigma_{02} = 1$$
$$\sigma_{03} = 1$$
$$\sigma_{11} = 1$$
$$\sigma_1 = 1.$$

The dual problem formulation is:

Objective Function $\quad \omega_{01} + \omega_{02} + \omega_{03} \quad\quad\quad = 1$ (13.3)
R terms $\quad\quad\quad 3\omega_{01} - \omega_{02} + \omega_{03} - \omega_{11} = 0$ (13.4)
L terms $\quad\quad\quad -2\omega_{01} \quad\quad\quad - 3\omega_{03} + 3\omega_{11} = 0.$ (13.5)

From the constraint equation there is only one term, so:

$$\omega_{10} = \omega_{11}. \tag{13.6}$$

This adds one additional equation but also one additional term, so the degrees of difficulty are equal to:

$$D = T - (N+1) = 4 - (2+1) = 1 \geq 0. \tag{13.7}$$

The dual has more variables than equations, and thus another equation is needed to solve for the dual variables. The relationships between the primal and dual variables will be used to determine an additional equation, and the equation typically is non-linear. The approach presented [2] is the "substitution approach" to obtain the additional equation needed. The "dimensional analysis approach" will be presented later in this Chapter. The relationships between the primal and dual which are:

$$C_{01} R^3 L^{-2} = \omega_{01} P \tag{13.8}$$
$$C_{02} R^{-1} = \omega_{02} P \tag{13.9}$$
$$C_{03} R L^{-3} = \omega_{03} P \tag{13.10}$$
$$C_{11} R^{-1} L^3 = (\omega_{11}/\omega_{10}). \tag{13.11}$$

13.2. PRIMAL AND DUAL FORMULATION OF JOURNAL BEARING DESIGN

Since $\omega_{10} = \omega_{11}$, Equation (13.11) can be used to relate the primal variables, that is:

$$R = C_{11} L^3 .\tag{13.12}$$

Using Equation (13.12) in Equation (13.9), one obtains:

$$\begin{aligned}P &= C_{02}/(\omega_{02} R)\\ &= [C_{02}/(C_{11}\omega_{02} L^3)] .\end{aligned}\tag{13.13}$$

Using Equation (13.10) with Equations (13.12) and (13.13), one obtains after reducing terms:

$$\begin{aligned}L^3 &= (C_{03} R)/(\omega_{03} P)\\ &= [(C_{02}/(C_{03} C_{11}^2)) * (\omega_{03}/\omega_{02})] .\end{aligned}\tag{13.14}$$

Now using Equation (13.8) and the values for R and L, one obtains:

$$(C_{01} C_{11}^4 L^{10})/C_{02} = \omega_{01}/\omega_{02} .\tag{13.15}$$

Now using Equation (13.14) in Equation (13.15) and reducing it, one can obtain:

$$(C_{01} C_{02}^{7/3})/(C_{03}^{10/3} C_{11}^{8/3}) = (\omega_{01} \omega_{02}^{7/3}/\omega_{03}^{10/3}) ,\tag{13.16}$$

or the form of

$$(C_{01}^3 C_{02}^7)/(C_{03}^{10} C_{11}^8) = (\omega_{01}^3 \omega_{02}^7/\omega_{03}^{10}) .\tag{13.17}$$

Now using Equations (13.3) to (13.5) to solve for the dual variables in terms of ω_{02}, one obtains:

$$\omega_{01} = (3/7)\omega_{02} \tag{13.18}$$
$$\omega_{03} = 1 - (10/7)^*\omega_{02} \tag{13.19}$$
$$\omega_{11} = 1 - (8/7)^*\omega_{02} .\tag{13.20}$$

Using Equations (13.18) to (13.20) in Equation (13.17), one can obtain:

$$(3/7\omega_{02})^3 (\omega_{02})^7/[1 - ((10/7) * \omega_{02})] = (C_{01}^3 C_{02}^7)/(C_{03}^{10} C_{11}^8) = A$$
or $\quad (\omega_{02}/(1 - ((10/7)\omega_{02})) = [A * (7/3)^3]^{1/10} = B$
or $\quad \omega_{02} = (7B/(7 + 10B)) .\tag{13.21}$

Thus, the remaining dual variables can be solved for as:

$$\omega_{01} = 3B/(7 + 10B) \tag{13.22}$$
$$\omega_{03} = 7/(7 + 10B) \tag{13.23}$$
$$\omega_{11} = (7 + 2B)/(7 + 10B) .\tag{13.24}$$

13. JOURNAL BEARING DESIGN CASE STUDY

Using Equations (13.10) and (13.12)

$$\begin{aligned}
P &= (C_{03}/\omega_{03})RL^{-3} \\
&= (C_{03}/\omega_{03})(C_{11}L^3)L^{-3} \\
&= C_{11}C_{03}/\omega_{03} \\
&= (C_{11}C_{03})(1 + (10/7)B) \\
&= (C_{11}C_{03})(1 + (10/7)[(7/3)^3 A]^{1/10}) \\
&= (C_{11}C_{03})(1 + (10/7)(7/3)^{3/10}[(C_{01}^3 C_{02}^7)/(C_{03}^{10} C_{11}^8)]^{1/10}) \\
&= C_{11}C_{03} + C_{11}C_{03}(10/7)(7/3)^{3/10}[(C_{01}^3 C_{02}^7)/(C_{03}^{10} C_{11}^8)]^{1/10} \\
&= C_{11}C_{03} + (10/7)[((7/3)(C_{01})]^{3/10} * (C_{02})^{7/10} * (C_{11})^{2/10}. \quad (13.25)
\end{aligned}$$

Using Equation (13.9) to solve for R, one obtains:

$$\begin{aligned}
R &= C_{02}/(\omega_{02}P) \\
&= C_{02}/[(7B/(7 + 10B)) * (C_{11}C_{03})(1 + (10/7)B)] \\
&= C_{02}/[(7B/(7 + 10B)) * (C_{11}C_{03})(7 + (10B)7)] \\
&= (C_{02}/(C_{11}C_{03}))/B \\
&= (C_{02}/(C_{11}C_{03}))/[(7/3)^3 * C_{01}^3 C_{02}^7/(C_{03}^{10} C_{11}^8)]^{1/10} \\
&= [(3/7) * (C_{02}/C_{01})]^{3/10} C_{11}^{-2/10}. \quad (13.26)
\end{aligned}$$

Using Equation (13.10) to solve for L, one obtains:

$$\begin{aligned}
L &= [C_{03}R/(\omega_{03}P)]^{1/3} \\
&= [[C_{03} * [(3/7) * (C_{02}/C_{01})]^{3/10} C_{11}^{-2/10}]/[\omega_{03} * C_{11}C_{03}/\omega_{03}]]^{1/3} \\
&= [(3/7)(C_{02}/C_{01})]^{1/10} * C_{11}^{-4/10}. \quad (13.27)
\end{aligned}$$

The equations for P, R, and L are general equations but are rather complex equations compared to the previous problems illustrated. The solution was based upon determining an additional equation from the primal-dual relationships, which was highly non-linear and resulted in rather complex expressions for the variables. The additional equation along with the dual variables was used in the equations relating the primal and dual to determine the final expressions for the variables. This frequently happens when the degrees of difficulty are greater than zero.

For this particular example problem where $C_{01} = 0.44$, $C_{02} = 10$, $C_{03} = 0.592$ and $C_{11} = 8.62$, the value for A and B are:

$$A = (C_{01}^3 C_{02}^7)/(C_{03}^{10} C_{11}^8) = [(0.44)^3 (10)^7]/[(0.592)^{10}(8.62)^8] = 5.285 \quad (13.28)$$
$$B = [A(7/3)^3]^{1/10} = [5.285(7/3)^3]^{1/10} = 1.523. \quad (13.29)$$

Now using the equations for the dual variables, Equations (13.21) to (13.24), one obtains

$$\begin{aligned}
\omega_{02} &= 7B/(7 + 10B) = 0.480 \\
\omega_{01} &= 3B/(7 + 10B) = 0.205 \\
\omega_{03} &= 7/(7 + 10B) = 0.315 \\
\omega_{11} &= (7 + 2B)/(7 + 10B) = 0.452.
\end{aligned}$$

13.3. DIMENSIONAL ANALYSIS TECHNIQUE FOR ADDITIONAL EQUATION

From Equation (13.25) the value of P can be found as:

$$P = C_{11}C_{03} + (10/7)[((7/3)(C_{01})]^{3/10} * (C_{02})^{7/10} * (C_{11})^{2/10}$$
$$= (8.62)(0.592) + (10/7)[((7/3)(0.44)]^{0.3}(10)^{0.7}(8.62)^{0.2}$$
$$= 5.10 + 11.10$$
$$= \$16.2 .$$

The primal variables can be determined from Equations (13.26) and (13.27) as:

$$R = [(3/7)(C_{02}/C_{01})]^{3/10} C_{11}^{-2/10}$$
$$= [(3/7)(10/0.44)]^{3/10} 8.62^{-2/10}$$
$$= 1.29 \text{ in}$$
$$L = [(3/7)(C_{02}/C_{01})]^{1/10} C_{11}^{-4/10}$$
$$= [(3/7)(10/0.44)]^{1/10} 8.62^{-4/10}$$
$$= 0.530 \text{ in} .$$

Now, if the values of R and L are used in Equation (13.1) for evaluating the primal, one obtains:

$$P = C_{01} R^3 L^{-2} + C_{02} R^{-1} + C_{03} R L^{-3}$$
$$= 0.44(1.29)^3 (0.53)^{-2} + 10(1.29)^{-1} + 0.592(1.29)(0.53)^{-3}$$
$$= 3.363 + 7.752 + 5.130$$
$$= \$16.2 .$$

As in the previous case studies, the value of the primal and dual objective functions are equivalent.

13.3 DIMENSIONAL ANALYSIS TECHNIQUE FOR ADDITIONAL EQUATION

It was difficult to determine the additional equation by repeated substitution and other methods can be used. One method is the technique of dimensional analysis and this will be demonstrated to obtain Equation (13.17). The dimensional analysis approach sets up the primal dual relations of Equations (13.8)-(13.11) and setting the primal variables on one side and the dual variables, constants, and objective function on the other side and giving the terms variable exponents as illustrated in Equation (13.30).

$$(R^3 L^{-2})^A (R^{-1})^B (RL^{-3})^C (R^{-1}L^3)^D = 1$$
$$= (\omega_{01} P/C_{01})^A (\omega_{02} P/C_{02})^B (\omega_{03} P/C_{03})^C (1/C_{11})^D . \qquad (13.30)$$

13. JOURNAL BEARING DESIGN CASE STUDY

One must balance the exponents to remove the primal variables (R, L) and the dual objective function (P). This is done by:

$$R \text{ values} \quad 3A - B + C - D = 0 \quad (13.31)$$
$$L \text{ Values} \quad -2A \quad -3C + 3D = 0 \quad (13.32)$$
$$P \text{ values} \quad A + B + C \quad = 0. \quad (13.33)$$

There are four variables and three equations, so one will find three variables in terms of the fourth variable. If one adds 3 times Equation (13.31) to Equation (13.32), one obtains:

$$A = 3/7B. \quad (13.34)$$

Using Equations (13.34) and (13.33), one can determine that:

$$B = -7/10C, \quad (13.35)$$

which results in:

$$A = -3/10C. \quad (13.36)$$

Using Equations (13.31), (13.35), and (13.36), one obtains that

$$D = 8/10C. \quad (13.37)$$

If one lets $C = 10$, the $A = -3$, $B = -7$ and $D = 8$. Using these values in Equation (13.30), one can obtain Equation (13.17), that is:

$$(C_{01}^3 C_{02}^7)/(C_{03}^{10} C_{11}^8) = (\omega_{01}^3 \ \omega_{02}^7/\omega_{03}^{10}). \quad (13.17)$$

Another solution method that is often used is the constrained derivative approach. This method has the dual equations rearranged in terms of one unknown dual variable and these are substituted into the dual objective function. The objective function is set into logarithmic form and differentiated with respect to the unknown dual variable, set to zero, and then solved for the unknown dual variable. The solved dual variable is used in the dual equations to obtain the values of the other dual variables. This technique is demonstrated in Chapters 16, 17, and 18.

13.4 EVALUATIVE QUESTIONS

1. Use the values of $C_{01} = 0.54$, $C_{02} = 10$, $C_{03} = 0.65$ and $C_{11} = 13.00$, determine the values of A and B, of the dual variables and the value of the objective function. Also determine the values of L and R, and use these to determine P.

2. Determine the sensitivity of the objective function and the primal variables of R and L by changing one of the constants by 20% (such as C_{01}).

3. Use dimensional analysis to develop the additional equation for the dual variables (ω values) in terms of the constants (C values) from the following data where D and H are primal variables and Y is the dual objective function.

$$C_{01} D^2 H = \omega_{01} Y$$
$$C_{02} D^3 = \omega_{02} Y$$
$$C_{11} D^{-1} = \omega_{11}/\omega_{10}$$
$$C_{12} H^{-1} = \omega_{12}/\omega_{10}$$
$$C_{13} D H^{-1} = \omega_{13}/\omega_{10}$$

[Answer $(C_{02}/C_{01})(C_{11}/C_{12}) = (\omega_{02}/\omega_{01})(\omega_{11}/\omega_{12})$].

REFERENCES

[1] Beightler, C.S., Lo, Ta-Chen, and Bylander, H.G., "Optimal Design by Geometric Programming," ASME, *Journal of Engineering for Industry*, 1970, pp. 191–196. 59

[2] Creese, R.C., "A Primal-Dual Solution Procedure for Geometric Programming," ASME, *Journal of Mechanical Design*, 1971. (Also as Paper No. 79-DET-78.) 60

CHAPTER 14

Metal Casting Hemispherical Top Cylindrical Side Riser Case Study

14.1 INTRODUCTION

The sphere is the shape which will give the longest solidification time, so a riser with a hemispheric shaped top should be a more efficient riser than a cylindrical riser and thus be more economical. However, the problem of designing this type of riser is more complex and has two degrees of difficulty. The case study considered is a side riser with a hemispherical top and a cylindrical bottom for ease of molding and to provide a better connection to the casting. The top hemispherical cap has the same diameter(D) as the cylinder.

14.2 PROBLEM FORMULATION

The volume of the riser is the sum of the cylinder part and the hemisphere part and can be written as:

$$\begin{aligned} V &= \text{cylindrical part} \quad + \text{Hemispherical part} \\ V &= \pi D^2 H/4 \quad\quad\quad + \pi D^3/12 \\ SA &= \pi D^2/4 + \pi DH \quad + \pi D^2/2 \\ &= 3/4\pi D^2 + \pi DH \,. \end{aligned} \quad (14.1)$$
$$(14.2)$$

The constraint for riser design is Chvorinov's Rule, which is

$$t = K(V/SA)^2 \,, \qquad (14.3)$$

where
- t = solidification time (minutes or seconds)
- K = solidification constant for molding material (minutes/in^2 or seconds/cm^2)
- V = volume (in^3 or cm^3)
- SA = cooling surface area (in^2 or cm^2).

An illustration of the hemispherical top side riser is shown in Figure 14.1 where the radius of the hemisphere is the same as the radius of the cylinder which is the diameter (D) divided by two.

14. METAL CASTING HEMISPHERICAL TOP CYLINDRICAL SIDE RISER

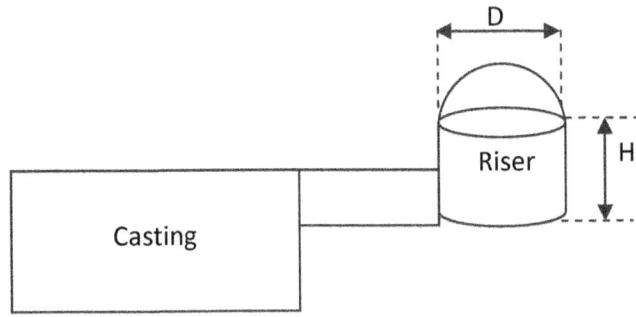

Figure 14.1: Hemispherical top side riser design.

This results in the relation:

$$(V/SA) \geq M_c = K, \tag{14.4}$$

where

M_c = the modulus of the casting (a constant K for a particular casting).

Thus, using the cooling surface area (SA) and volume (V) expressions, Equation (14.4) can be rewritten as:

$$\begin{aligned} V/SA &= (\pi D^2 H/4 + \pi D^3/12)/(3/4\pi D^2 + \pi DH) & &\geq K \\ &= (\pi D^2/12) * (3H + D)/[(\pi D/4) * (3D + 4H)] & &\geq K \\ &= D * (3H + D)/[3 * (3D + 4H)] & &\geq K. \end{aligned} \tag{14.5}$$

Rearranging the equation in the less than equal form results in:

$$4KD^{-1} + 3KH^{-1} - (1/3)DH^{-1} \leq 1. \tag{14.6}$$

Thus, the primal form of the problem can be stated as:

$$\text{Min } V = \pi D^2 H/4 + \pi D^3/12. \tag{14.7}$$

Subject to:

$$4KD^{-1} + 3KH^{-1} - (1/3)DH^{-1} \leq 1. \tag{14.8}$$

From the coefficients and signs, the signum values for the dual are:

$$\begin{aligned} \sigma_{01} &= 1 \\ \sigma_{02} &= 1 \\ \sigma_{11} &= 1 \\ \sigma_{12} &= 1 \\ \sigma_{13} &= -1 \\ \sigma_1 &= 1. \end{aligned}$$

14.2. PROBLEM FORMULATION

The dual problem formulation is:

Objective Function	$\omega_{01} + \omega_{02}$	$= 1$	(14.9)
D terms	$2\omega_{01} + 3\omega_{02} - \omega_{11} - \omega_{13}$	$= 0$	(14.10)
H terms	$\omega_{01} - \omega_{12} + \omega_{13}$	$= 0$.	(14.11)

The degrees of difficulty are equal to:

$$D = T - (N+1) = 5 - (2+1) = 2. \tag{14.12}$$

Using the linearity inequality equation,

$$\omega_{10} = \omega_{mt} = \sigma_m \sum \sigma_{mt}\omega_{mt} = (1) * (1 * \omega_{11} + 1 * \omega_{12} + (-1) * \omega_{13})$$

$$\omega_{10} = \omega_{11} + \omega_{12} - \omega_{13}. \tag{14.13}$$

The dual variables cannot be determined directly as the degrees of difficulty are 2; that is, there are two more variables than there are equations. The objective function can be found using the dual expression:

$$V = d(\omega) = \sigma \left[\prod_{m=0}^{M} \prod_{t=1}^{T_m} (C_{mt}\omega_{mo}/\omega_{mt})^{\sigma_{mt}\omega_{mt}} \right]^{\sigma} \tag{14.14}$$

$$V = 1[[\{(\pi/4) * 1/\omega_{01})\}(1 * \omega_{01})] * [\{(\pi/12) * \omega_{02})\}^{(1*\omega_{02})}] * [\{(4K * \omega_{10}/\omega_{11})\}^{(1*\omega_{10})}] *$$
$$[\{(3K * \omega_{10}/\omega_{12})\}^{(1*\omega_{10})}] * [\{((1/3) * \omega_{10}/\omega_{13})\}^{(-1*\omega_{13})}]]^{1}. \tag{14.15}$$

The relationships between the primal and dual variables can be written as:

$$(\pi/4)KD^2H = \omega_{01}V \tag{14.16}$$
$$(\pi/12)KD^3 = \omega_{02}V \tag{14.17}$$
$$(4K)D^{-1} = \omega_{11}/\omega_{10} \tag{14.18}$$
$$(3K)H^{-1} = \omega_{12}/\omega_{10} \tag{14.19}$$
$$(1/3)DH^{-1} = \omega_{13}/\omega_{10}. \tag{14.20}$$

If one takes Equation (14.16) and divides by Equation (14.17), one obtains:

$$3H/D = \omega_{01}/\omega_{02}. \tag{14.21}$$

If one takes the inverse of Equation 10.20, one obtains:

$$3H/D = \omega_{10}/\omega_{13}. \tag{14.22}$$

Now comparing Equations (14.21) and (14.22), one can obtain an equation between the dual variables as:

$$\omega_{01}/\omega_{02} = \omega_{10}/\omega_{13}. \tag{14.23}$$

14. METAL CASTING HEMISPHERICAL TOP CYLINDRICAL SIDE RISER

If one takes Equation (14.18) and divides by Equation (14.19), one obtains:

$$(4/3)H/D = \omega_{11}/\omega_{12} . \tag{14.24}$$

Now comparing Equations (14.21) and (14.24), one can obtain an additional equation between the dual variables as:

$$\omega_{01}/\omega_{02} = (9/4)\omega_{11}/\omega_{12} . \tag{14.25}$$

Now there are six equations with only the six dual variables, and they are Equations (14.9) to (14.11), (14.13), (14.23), and (14.25). The procedure used was to solve for all of the variables in terms of ω_{02} and then obtain the specific value of ω_{02}.

From Equation (14.9), one obtains:

$$\omega_{01} = 1 - \omega_{02} . \tag{14.26}$$

If one adds Equations (14.10) and (14.11), one obtains:

$$3\omega_{01} + 3\omega_{02} - \omega_{11} - \omega_{12} = 0 .$$

Which can be reduced to:

$$\omega_{11} + \omega_{12} = 3 . \tag{14.27}$$

Using Equations (14.13) and (14.27), one obtains:

$$\omega_{10} = \omega_{11} + \omega_{12} - \omega_{13}$$
$$\omega_{10} = 3 - \omega_{13} . \tag{14.28}$$

Now using Equations (14.28) and (14.23), one obtains:

$$\omega_{01}/\omega_{02} = \omega_{10}/\omega_{13} = (3 - \omega_{13})/\omega_{13} = (1 - \omega_{02})/\omega_{02} .$$

Solving for ω_{13} one obtains:

$$\omega_{13} = 3\omega_{02} . \tag{14.29}$$

From Equations (14.28) and (14.29):

$$\omega_{10} = 3 - \omega_{13} = 3 - 3\omega_{02}$$
$$\omega_{10} = 3(1 - \omega_{02}) . \tag{14.30}$$

Using Equations (14.26) and (14.29) in Equation (14.11), one obtains:

$$\omega_{12} = \omega_{10} + \omega_{13}$$
$$= (1 - \omega_{02}) + 3\omega_{02}$$
$$= 1 + 2\omega_{02} . \tag{14.31}$$

14.2. PROBLEM FORMULATION

Using Equations (14.10), (14.26), and (14.29), one has:

$$\begin{aligned}\omega_{11} &= 2\omega_{01} + 3\omega_{02} - \omega_{13} \\ &= 2(1-\omega_{02}) + 3\omega_{02} - 3\omega_{02} \\ &= 2(1-\omega_{02}) \,.\end{aligned} \quad (14.32)$$

Now using Equation (14.25), one can solve for ω_{02} using the values for ω_{01}, ω_{11}, and ω_{12}

$$\omega_{01}/\omega_{02} = (9/4)\omega_{11}/\omega_{12}$$
$$(1-\omega_{02})/\omega_{02} = (9/4)2(1-\omega_{02})/(1+2\omega_{02}) \,.$$

And solving for ω_{02} results in:

$$\omega_{02} = 0.4 \,.$$

Therefore,

$\omega_{01} = 0.6$
$\omega_{11} = 1.2$
$\omega_{12} = 1.8$
$\omega_{13} = 1.2$
$\omega_{10} = 1.8 \,.$

Now using the dual variables in Equation (14.15) to find the minimum volume, one obtains:

$$\begin{aligned}V &= 1[[\{(\pi/4*1/\omega_{01})\}^{(1*\omega_{01})}] * [\{(\pi/12*\omega_{02})\}^{(1*\omega_{02})}] * [\{(4K*\omega_{10}/\omega_{11})\}^{(1*\omega_{11})}] * \\ & \quad [\{(3K*\omega_{10}/\omega_{12})\}^{(1*\omega_{12})}] * [\{((1/3)*\omega_{10}/\omega_{13})\}^{(-1*\omega_{13})}]]^1 \\ &= 1[[\{(\pi/4)*1/0.6)\}^{(1*0.6)}] * [\{(\pi/12*0.4)\}^{(1*0.4)}] * [\{(4K*1.8/1.2)\}^{(1*1.2)}] * \\ & \quad [\{(3K*1.8/1.8)\}^{(1*1.8)}] * [\{((1/3)*1.8/1.2)\}^{(-1*1.2)}]]^1 \\ &= 1[[\{((5/12)\pi)\}^{(0.6)}] * [\{((5/24)\pi)\}^{(0.4)}] * [\{(6K)\}^{(1.2)}] * [\{(3K)\}^{(1.8)}] * [\{(1/2)\}^{(-1.2)}]]^1 \\ &= 1[[\{((2*5/24)\pi)\}^{(0.6)}] * [\{((5/24)\pi)\}^{(0.4)}] * [\{(2*3K)\}^{(1.2)}] * [\{(3K)\}^{(1.8)}] * [\{(2)\}^{(1.2)}]]^1 \\ &= 2^{0.6}[(5/24)\pi]^{(0.6+0.4)} * 2^{1.2} * (3K)^{(1.2+1.8)}2^{1.2} \\ &= (5\pi/24)^1 * 2^{(0.6+1.2+1.2)} * (3K)^3 \\ V &= (5\pi/24) * (6K)^3 \,.\end{aligned} \quad (14.33)$$

The values for H and D can be found from Equations (14.18) and (14.19)

$$\begin{aligned}D &= 4K*(\omega_{10}/\omega_{11}) \\ &= 4k*(1.8/1.2) \\ D &= 6K\end{aligned} \quad (14.34)$$

and

$$\begin{aligned}H &= 3K*(\omega_{10}/\omega_{12}) \\ &= 3K*(1.8/1.8) \\ H &= 3K \,.\end{aligned} \quad (14.35)$$

Now the primal can be evaluated using Equation (14.7) with the values for H and D from Equations (14.34) and (14.35).

$$V = \pi D^2 H/4 + \pi D^3/12$$
$$V = \pi(6K)^2(3K)/4 + \pi(6K)^3/12$$
$$= \pi 27K^3 + \pi 18K^3$$
$$= 45\pi K^3 \tag{14.36}$$
$$= (5\pi/24) * (6K)^3 . \tag{14.37}$$

The values for the primal and dual are equivalent, which is required for the solution. The expression of Equation (14.37) is the preferred expression for foundry as $6K$ is the value for the diameter for the simple cylindrical risers where K is the modulus of the casting. This example illustrates that it is possible to solve problems with two degrees of difficulty in some instances, but there are numerous mathematical operations that must be performed to obtain the additional equations and solution. There have been several articles written concerning various riser design shapes and the use of insulating materials to improve casting yield [1, 2, 3, 4, 5].

14.3 DIMENSIONAL ANALYSIS TECHNIQUE FOR ADDITIONAL TWO EQUATIONS

Although it was relatively easy to determine the two additional equations, they also could have been obtained by dimensional analysis. The dimensional analysis approach sets up the primal dual relations of Equations (14.16)-(14.20) and setting the primal variables on one side and the dual variables, constants, and objective function on the other side and giving the terms variable exponents as illustrated in Equation (14.38)

$$(D^2H)^A(D^3)^B(D^{-1})^C(H^{-1})^D(DH^{-1})^E = 1$$
$$= (\omega_{01}V/C_{01})^A(\omega_{02}V/C_{02})^B(\omega_{11}/\omega_{10}C_{11})^C(\omega_{12}/\omega_{10}C_{12})^D(\omega_{13}/\omega_{10}C_{13})^E , \tag{14.38}$$

where

$$C_{01} = (K\pi/4) \tag{14.39}$$
$$C_{02} = (K\pi/12) \tag{14.40}$$
$$C_{11} = 4K \tag{14.41}$$
$$C_{12} = 3K \tag{14.42}$$
$$C_{13} = 1/3 . \tag{14.43}$$

14.3. DIMENSIONAL ANALYSIS TECHNIQUE FOR ADDITIONAL TWO EQUATIONS

One must balance the exponents to remove the primal variables (DH) and the dual objective function (V). An equation will also be written to consider ω_{10}. This is done by:

$$V \text{ values} \quad A + B = 0 \tag{14.44}$$
$$\omega_{10} \text{ values} \quad +C + D + E = 0 \tag{14.45}$$
$$D \text{ values} \quad 2A + 3B - C + E = 0 \tag{14.46}$$
$$H \text{ values} \quad A - D - E = 0. \tag{14.47}$$

There are five variables and four equations, so one will find three variables in terms of the fourth variable. From Equation (14.44), one notes that

$$A = -B. \tag{14.48}$$

If Equations (14.46) and (14.47) are added and using the relation of (14.48), one obtains

$$C = -D. \tag{14.49}$$

Using the relationship of (14.49) and (14.45), one notes that

$$E = 0. \tag{14.50}$$

If one takes Equation (14.46) and subtracts two times Equation (14.44) and using $E = 0$, one obtains that

$$C = B. \tag{14.51}$$

Thus, if one lets $B = 1$, then $C = 1$, $D = -1$, $A = -1$ and $E = 0$. Using these values in (14.38) one obtains that

$$(\omega_{02}/\omega_{01})(\omega_{11}/\omega_{12}) = (C_{02}/C_{01})(C_{11}/C_{12})$$
$$= ((K\pi/12)/(K\pi/4))((4K)/(3K)) = 4/9. \tag{14.52}$$

Upon rearranging terms, this is equivalent to Equation (14.25).

The second equation is obtained by allowing ω_{10} to be part of the new equation. Thus, taking the Equations (14.44)-(14.47) and allow ω_{10} to enter the solution by making the $RHS = 1$ in Equation (14.45), which is renumbered as (14.53).

$$V \text{ values} \quad A + B = 0 \tag{14.44}$$
$$\omega_{10} \text{ values} \quad +C + D + E = 1 \tag{14.53}$$
$$D \text{ values} \quad 2A + 3B - C + E = 0 \tag{14.46}$$
$$H \text{ values} \quad A - D - E = 0. \tag{14.47}$$

There are five variables and four equations, so one will find three variables in terms of the fourth variable. From Equation (14.44), one notes that

$$A = -B. \tag{14.48}$$

If Equations (14.46) and (14.47) are added and using the relation of (14.48), one obtains

$$C = -D. \tag{14.49}$$

Using the relationship of (14.49) and (14.52), one notes that

$$E = 1. \tag{14.53}$$

Using Equations (14.47) and (14.53), one obtains that

$$A = D + 1 \tag{14.54}$$

and thus

$$B = -(D+1). \tag{14.55}$$

Using Equation (14.46) and (14.48), one obtains that

$$C = B + E = -D. \tag{14.56}$$

Thus, if one lets $D = 1$, then $A = 2$, $B = -2$, $C = -1$ and $E = 1$. Using these values in (14.38) one obtains that

$$(\omega_{01}V/C_{01})^2(\omega_{02}V/C_{02})^{-2}(\omega_{11}/\omega_{10}C_{11})^{-1}(\omega_{12}/\omega_{10}C_{12})^1(\omega_{13}/\omega_{10}C_{13})^1 = 1.$$

Reducing terms and putting the constants on the right-hand side

$$(\omega_{01}/\omega_{02})^2(\omega_{12}/\omega_{11})(\omega_{13}/\omega_{10}) = (C_{12}C_{13}C_{01}^2)/(C_{02}^2 C_{11}) = (3K * 1/3 * (K\pi/4)^2 / ((K\pi/12)^2 4K)$$
$$(\omega_{01}/\omega_{02})^2(\omega_{12}/\omega_{11})(\omega_{13}/\omega_{10}) = 9/4. \tag{14.57}$$

But rearranging Equation (14.25), one observes that

$$(\omega_{01}/\omega_{02})(\omega_{12}/\omega_{11}) = 9/4.$$

Cancelling these values from both sides, the result is:

$$(\omega_{01}/\omega_{02})(\omega_{13}/\omega_{10}) = 1, \tag{14.58}$$

which is equivalent to Equation (14.23).

$$\omega_{01}/\omega_{02} = \omega_{10}/\omega_{13}. \tag{14.23}$$

Thus, the dimensional analysis approach gives the same results as the substitution approach for obtaining the additional equations. The advantage of the dimensional analysis approach is that it is more straight forward in obtaining the relationships.

14.4 EVALUATIVE QUESTIONS

1. A side riser with a hemispheric top is to be designed for a casting which has a surface area of 40 cm^2 and a volume of 120 cm^3. The hot metal cost is 100 Rupees per kg and the metal density is 3.0 gm/cm^3. Compare these results with Problem 1 in Section 6.2.

    (a) What are the dimensions of the hemispherical side riser (H and D)?

    (b) What is the volume of the hemispherical side riser (cm^3)?

    (c) What is the metal cost of the hemispherical side riser (Rupees)?

    (d) What is the metal cost of the casting (Rupees)?

2. Two castings of equal volume but of different dimensions are to be cast. If one is a 3 inch cube and the other is a plate of $1 \times 3 \times 9$ inches and a top riser is to be used, what are the dimensions (H and D) of the risers for the two cases?

REFERENCES

[1] Creese, R. C., Optimal Riser Design by Geometric Programming," *AFS Cast Metals Research Journal,* Vol. 7, 1971, pp. 118–121. 72

[2] Creese, R.C., "Dimensioning of Risers for Long Freezing Range Alloys by Geometric Programming," *AFS Cast Metals Research Journal,* Vol. 7, 1971, pp. 182–184. 72

[3] Creese, R.C., "Generalized Riser Design by Geometric Programming," *AFS Transactions,* Vol. 87, 1979, pp. 661–664. 72

[4] Creese, R.C., "An Evaluation of Cylindrical Riser Designs with Insulating Materials," *AFS Transactions,* Vol. 87, 1979, pp. 665–669. 72

[5] Creese, R.C., "Cylindrical Top Riser Design Relationships for Evaluating Insulating Materials," *AFS Transactions,* Vol. 89, 1981, pp. 345–348. 72

CHAPTER 15

Liquefied Petroleum Gas (LPG) Cylinders Case Study

15.1 INTRODUCTION

This case study problem [1, 2] deals with the design of liquefied petroleum gas cylinders, more commonly known as propane gas cylinders in the USA. This is a very interesting problem as it has two general solutions as well as one degree of difficulty. The two general solutions occur depending upon the relationship between the constants. What happens in this particular problem is that one of the two constraints can be either binding or loose, depending upon the value of the constants.

15.2 PROBLEM FORMULATION

The problem was to minimize the drawing force (Z) to produce the tank by deep drawing and two constraints were considered so the tank would have a minimum volume and the height/diameter ratio would be less than one. The formulation of the primal problem was:

$$\text{Minimize } Z = K_1 hd + K_2 d^2 , \tag{15.1}$$

subject to the two constraints:

$$\pi d^2 h/4 \geq V_{\min} , \tag{15.2}$$

or in the proper geometric programming form as

$$(4V_{\min}/\pi)d^{-2}h^{-1} \leq 1 , \tag{15.3}$$

and

$$h/d \leq 1 , \tag{15.4}$$

where

$$K_1 = \pi PYC/F , \text{ and} \tag{15.5}$$
$$K_2 = ((C - E)\pi PY/2F , \tag{15.6}$$

where

Z = drawing force
P = internal gas pressure
Y = material yield strength
F = hoop stress
C = constant = 1.04
E = constant = 0.65.

If all the constants are combined, the primal form can be written as:

$$Z = K_1 hd + K_2 d^2, \tag{15.1}$$

subject to:

$$K_3 h^{-1} d^{-2} \leq 1, \tag{15.7}$$

and

$$K_4 hd^{-1} \leq 1, \tag{15.8}$$

where

$$K_3 = (4V_{\min}/\pi), \tag{15.9}$$

and

$$K_4 = 1 \quad \text{(This could be taken as the minimum } d/h \text{ ratio).} \tag{15.10}$$

From the coefficients and signs, the signum values for the dual are:

$$\sigma_{01} = 1$$
$$\sigma_{02} = 1$$
$$\sigma_{11} = 1$$
$$\sigma_{21} = 1$$
$$\sigma_1 = 1$$
$$\sigma_2 = 1.$$

The dual problem formulation is:

Objective Function	$\omega_{01} + \omega_{02}$	$= 1$	(15.11)
h terms	$\omega_{01} - \omega_{11} + \omega_{21} = 0$		(15.12)
d terms	$\omega_{01} + 2\omega_{02} - 2\omega_{11} - \omega_{21} = 0.$		(15.13)

The degrees of difficulty are equal to:

$$D = T - (N+1) = 4 - (2+1) = 1. \tag{15.14}$$

From the constraint equations which have only one term it is apparent that:

$$\omega_{10} = \omega_{11}, \tag{15.15}$$

15.2. PROBLEM FORMULATION

and
$$\omega_{20} = \omega_{21} . \tag{15.16}$$

An additional equation is needed, and one must examine the primal dual relationships to find the additional relationship, which are:

$$K_1 hd = \omega_{01} Z \tag{15.17}$$
$$K_2 d^2 = \omega_{02} Z \tag{15.18}$$
$$K_3 h^{-1} d^{-2} = \omega_{11}/\omega_{10} = 1 \tag{15.19}$$
$$K_4 hd^{-1} = \omega_{21}/\omega_{20} = 1 . \tag{15.20}$$

If one adds Equations (15.12) and (15.13), one can solve for ω_{11} by:

$$2\omega_{01} + 2\omega_{02} - 3\omega_{11} = 0 \tag{15.21}$$
$$2(\omega_{01} + \omega_{02}) - 3\omega_{11} = 0 \tag{15.22}$$
$$2 - 3\omega_{11} = 0 \tag{15.23}$$
$$\omega_{11} = 2/3 . \tag{15.24}$$

If one takes Equation (15.17) and divides it by Equations (15.18) and (15.20), one obtains:

$$K_1 hd/(K_2 d^2 * K_4 hd^{-1}) = K_1/(K_2 K_4) = \omega_{01} Z/(\omega_{02} Z * 1) = \omega_{01}/\omega_{02} . \tag{15.25}$$

This can be solved for ω_{01} in terms of ω_{02} and the constants:

$$\omega_{01} = \omega_{02}(K_1/(K_2 K_4)) . \tag{15.26}$$

Now using Equations (15.26) and (15.11), one can solve for ω_{01} and ω_{02} and obtain:

$$\omega_{01} = K_1/(K_1 + K_2 K_4) \tag{15.27}$$
$$\omega_{02} = K_2 K_4/(K_1 + K_2 K_4) . \tag{15.28}$$

Now using Equation (15.12) and substituting the values for ω_{01} and ω_{11}, one obtains:

$$\omega_{21} = \omega_{11} - \omega_{01} = (2K_2 K_4 - K_1)/[3(K_1 + K_2 K_4)] . \tag{15.29}$$

Now ω_{21} must be ≥ 0, so that implies that:

$$2K_2 K_4 - K_1 \geq 0 , \tag{15.30}$$

or

$$K_2 K_4/K_1 \geq 1/2 . \tag{15.31}$$

Thus, there are two sets of solutions, depending upon whether Equation (15.31) holds; this can be illustrated in Figure 15.1.

Now, the solutions can be found for the two cases. If the answer is "No," then the objective function can be evaluated using the dual expression:

$$Z = d(\omega) = \sigma \left[\prod_{m=0}^{M} \prod_{t=1}^{T_m} (C_{mt}\omega_{mo}/\omega_{mt})^{\sigma_{mt}\omega_{mt}} \right]^{\sigma}, \quad (15.32)$$

where by definition

$$\omega_{00} = 1. \quad (15.33)$$

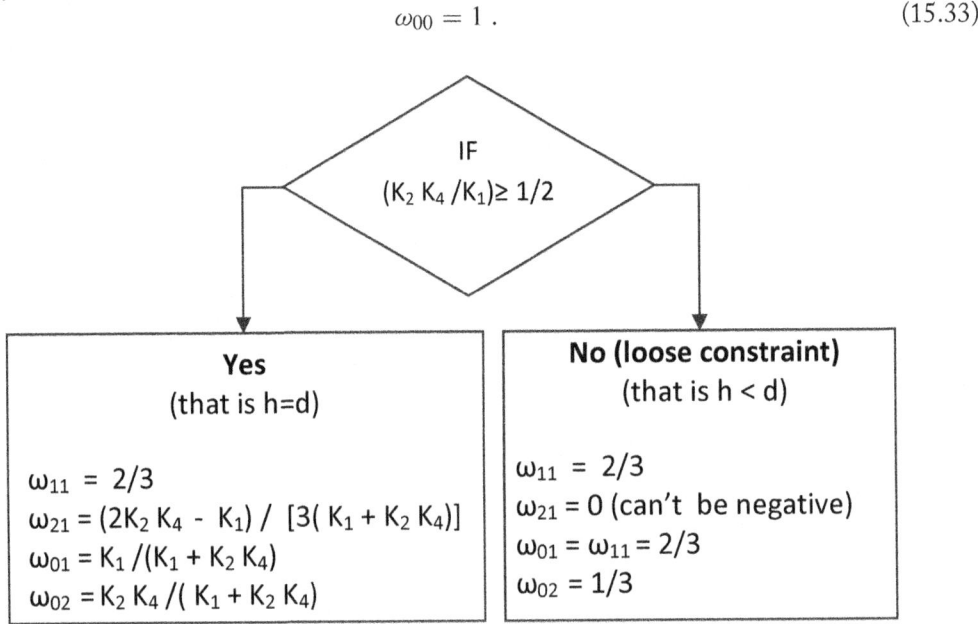

Figure 15.1: Values of Dual Variables based upon making $\omega_{21} \geq 0$.

$$Z = [K_1 * 1/(2/3)]^{1*2/3}[K_2 * 1/(1/3)]^{1*1/3}[K_3 * (2/3)/(2/3)]^{1*2/3}, \quad (15.34)$$

which can be reduced to:

$$Z = [3/2^{2/3}]K_1^{2/3}K_2^{1/3}K_3^{2/3}. \quad (15.35)$$

Now, the primal variables can be determined from the primal dual relationships. If one uses Equation (15.18) and solves for d^2 and then for d, one obtains:

$$d^2 = \omega_{02}Z/K_2 = (1/3) * \{[3/2^{2/3}]K_1^{2/3}K_2^{1/3}K_3^{2/3}\}/K_2 = (K_1K_3/2K_2)^{2/3}. \quad (15.36)$$

Solving for d results in:

$$d = (K_1K_3/2K_2)^{1/3}. \quad (15.37)$$

15.2. PROBLEM FORMULATION

Similarly, if one uses Equation (15.17) to solve for h, one obtains:

$$h = \omega_{01} Z / K_1 d = \{(2/3) * \{[3/2^{2/3}] K_1^{2/3} K_2^{1/3} K_3^{2/3}\}\}/[K_1 * (K_1 K_3/2 K_2)^{1/3}]. \quad (15.38)$$

Reducing terms, one obtains:

$$h = 2^{2/3} K_1^{-2/3} K_2^{2/3} K_3^{1/3}. \quad (15.39)$$

Now substituting the primal variables into the primal objective function, one has:

$$\begin{aligned}
Z &= K_1 h d + K_2 d^2, & (15.1)\\
&= K_1 * 2^{2/3} K_1^{-2/3} K_2^{2/3} K_3^{1/3} * (K_1 K_3/2K_2)^{1/3} + K_2 * (K_1 K_3/2K_2)^{2/3}\\
&= 2 K_1^{2/3} K_2^{1/3} K_3^{2/3}/2^{2/3} + K_1^{2/3} K_2^{1/3} K_3^{2/3}/2^{2/3}\\
&= 3 K_1^{2/3} K_2^{1/3} K_3^{2/3}/2^{2/3}\\
&= [3/2^{2/3}] * K_1^{2/3} K_2^{1/3} K_3^{2/3}. & (15.40)
\end{aligned}$$

The equations for the primal and dual, Equations (15.35) and (15.40), give the same results. Thus, one has a general solution for the objective function and the two primal variables when the "No" route was taken.

The "Yes" route has the dual variables as functions of the constants and thus is the more complex route. The objective function can be evaluated using the dual expression:

$$Z = d(\omega) = \sigma \left[\prod_{m=0}^{M} \prod_{t=1}^{T_m} (C_{mt} \omega_{mo}/\omega_{mt})^{\sigma_{mt} \omega_{mt}} \right]^{\sigma}, \quad (15.32)$$

where by definition

$$\omega_{00} = 1. \quad (15.33)$$

The dual variables are:

$$\begin{aligned}
\omega_{11} &= 2/3 & (15.41)\\
\omega_{21} &= (2K_2 K_4 - K_1)/[3(K_1 + K_2 K_4)] & (15.42)\\
\omega_{01} &= K_1/(K_1 + K_2 K_4) & (15.43)\\
\omega_{02} &= K_2 K_4/(K_1 + K_2 K_4). & (15.44)
\end{aligned}$$

Using these dual variables and the signum values the dual objective function is:

$$\begin{aligned}
Z &= (K_1 \omega_{00}/\omega_{01})^{1*\omega_{01}} (K_2 \omega_{00}/\omega_{02})^{1*\omega_{02}} (K_3)^{1*\omega_{11}} (K_4)^{1*\omega_{21}}\\
&= (K_1 + K_2 K_4)^{\omega_{01}} [(K_1 + K_2 K_4)/K_4]^{\omega_{02}} (K_3)^{2/3} (K_4)^{\omega_{11} - \omega_{01}}\\
&= [(K_1 + K_2 K_4)/K_4]^{\omega_{01}} [(K_1 + K_2 K_4)/K_4]^{\omega_{02}} (K_3 K_4)^{2/3}\\
&= [(K_1 + K_2 K_4)/K_4] (K_3 K_4)^{2/3}\\
&= (K_3)^{2/3} [(K_1 K_4^{-1/3} + K_2 K_4^{2/3})]. & (15.45)
\end{aligned}$$

The primal variables can be obtained from the relationships between the primal and dual variables. Using Equation (15.20), one can obtain a relation between h and d which is:

$$h = d/K_4 . \tag{15.46}$$

If one combines Equations (15.19) and (15.20) one obtains:

$$(K_3 h^{-1} d^{-2})(K_4 h d^{-1}) = 1 * 1 ,$$

which yields:

$$K_3 K_4 d^{-3} = 1 ,$$

or

$$d = (K_3 K_4)^{1/3} . \tag{15.47}$$

Now from Equations (15.46) and (15.47) one obtains:

$$h = d/K_4 = K_3^{1/3} K_4^{-2/3} . \tag{15.48}$$

Now, using the primal equation for the objective function with the primal variables, one has:

$$\begin{aligned} Z &= K_1 h d + K_2 d^2 , \tag{15.1}\\ &= K_1 (K_3^{1/3} K_4^{-2/3})(K_3 K_4)^{1/3} + K_2 (K_3 K_4)^{2/3} \\ &= K_1 (K_3^{2/3} K_4^{-1/3}) + K_2 (K_3 K_4)^{2/3} \\ &= K_3^{2/3} (K_1 K_4^{-1/3} + K_2 K_4^{2/3}) . \tag{15.49} \end{aligned}$$

Note that the general objective function is the same for both the primal and dual solutions. This problem illustrates that it is possible to solve a problem with more than one degree of difficulty and have two solutions, depending upon the specific values of the constants in the problem.

15.3 DIMENSIONAL ANALYSIS TECHNIQUE FOR ADDITIONAL EQUATION

It was not very difficult to determine the additional equation by repeated substitution, but the technique of dimensional analysis can be used to verify it and this will be demonstrated to obtain Equation (15.26). The (15.17)– (15.20) and setting the primal variables on one side and the dual variables, constants, and objective function on the other side and giving the terms variable exponents as illustrated in Equation (15.50)

$$(hd)^A (d^2)^B (h^{-1} d^{-2})^C (hd^{-1})^D = 1 = (\omega_{01} Z/K_1)^A (\omega_{02} Z/K_2)^B (1/K_3)^C (1/K_4)^D . \tag{15.50}$$

One must balance the exponents to remove the primal variables (d, h) and the dual objective function (Z). This is done by:

$$\begin{array}{lll} Z \text{ values} & A + B & = 0 \tag{15.51}\\ h \text{ values} & A \quad\quad - C + D = 0 \tag{15.52}\\ d \text{ values} & A + 2B - 2C - D = 0 . \tag{15.53} \end{array}$$

15.4. EVALUATIVE QUESTIONS

There are four variables and three equations, so one will find three variables in terms of the fourth variable. From Equation (15.51), one obtains

$$A = -B \,. \tag{15.54}$$

Using Equations (15.54) and (15.52), one notes that

$$B = D - C \,. \tag{15.55}$$

If one subtracts Equation (15.52) from Equation (15.53), one obtains that

$$B = D + C/2 \,. \tag{15.56}$$

If one compares Equations (15.55) and (15.56), the only solution was for both equations to be satisfied is if:

$$C = 0 \,. \tag{15.57}$$

If one sets $D = 1$, then $B = 1$ and $A = -1$. Using these values in Equation (15.50) one obtains

$$(hd)^{-1}(d^2)^1(h^{-1}d^{-2})^0(hd^{-1})^1 = 1 = (\omega_{01}Z/K_1)^{-1}(\omega_{02}Z/K_2)^1(1/K_3)^0(1/K_4)^1 \,. \tag{15.58}$$

This results in Equation (15.59) which is equivalent to Equation (15.26)

$$\omega_{01} = \omega_{02}(K_1/(K_2 K_4)) \,. \tag{15.59}$$

The solution to the problem is the same, but the obtaining of the additional equation is often the difficult step in the process. The dimensional analysis technique is often easier than the substitution method as it is more consistent and often faster in obtaining the necessary additional equation(s).

15.4 EVALUATIVE QUESTIONS

1. A tank is to be designed with a minimum volume of 17,500,000 mm³ and the values for parameters are:

   $P = 0.2535$ kg/mm²
   $F = 32.33$ kg/mm²
   $Y = 25$ kg/mm²
   $C = 1.04$
   $E = 0.65$.

   Determine the amount of the drawing force (kg) and the height (mm) and diameter (mm) of the tank.

2. A new procedure was developed for another older machine which changed the expression for K_2. The new expression was:

$$K_2 = (2C - E)\pi PY/(2F). \tag{15.50}$$

Using the same data as in Evaluative Question 1, determine the amount of the drawing force (kg) and the height (mm) and diameter (mm) of the new tank.

3. A tank is to be designed with a minimum volume of 11,000 in^3 and the values for parameters are:
   $P = 360$ lb/in^2
   $F = 46,000$ lb/in^2
   $Y = 35,500$ lb/in^2
   $C = 1.04$
   $E = 0.65$ $\omega_{01} = \omega_{02}(K_1/(K_2 K_4))$.

   Determine the amount of the drawing force (lb and tons) and the height (in) and diameter (in) of the tank.

REFERENCES

[1] Ravivarna Pericherla, *Design and Manufacture of Liquefied Petroleum Gas Cylinders,* MS Thesis, West Virginia University, 1992, Morgantown, WV, USA. 77

[2] B. Gopalakrishnan and R. Pericherla, "Design for Manufacturability of the Liquefied Petroleum Gas Cylinder," *Proceedings of the International Industrial Engineering Research Conference,* IIE, Miami, FL, May 1997. 77

CHAPTER 16

Material Removal/Metal Cutting Economics with Two Constraints

16.1 INTRODUCTION

The metal cutting economics case study in Chapter 12 had only a feed constraint, but the problem presented here also includes the horsepower constraint. The additional constraint leads to a problem with one degree of difficulty and to the possibility of multiple solutions similar to that in the Liquefied Petroleum Gas Cylinder Case Study. Thus, the problem is more difficult to solve and needs to be considered separately, but the equations to define the problem are similar to those of Chapter 12. The theory is from the work by Tsai [1, 3] and the example data is from that of Ermer [2].

16.2 PROBLEM FORMULATION

The problem is based upon the modified form of the Taylor's Tool Life Equation and two constraints, the feed rate and horsepower constraints. The cost function represents the sum of the various costs and is expressed as:

$$C_u = (R_o + R_m)t_h + (R_o + R_m)t_c + C_t n_t + (R_o + R_m)t_{ch}n_t , \qquad (16.1)$$

where

C_u = Unit Cost, \$/piece
R_o = Operator Rate, \$/min
R_m = Machine Rate, \$/min
C_t = Tool Cost (per cutting edge for insert tools), \$
t_h = Handling Time (to insert and remove piece), min/piece
t_c = Cutting Time, min/piece
t_{ch} = Tool Changing Time (to change tool), min
n_t = Number of tool changes per piece.

The modified form of Taylor's Tool life Equation is:

$$TV^{1/n} f^{1/m} = C \qquad (16.2)$$

where

16. MATERIAL REMOVAL/METAL CUTTING ECONOMICS

T = Tool Life, min
V = Cutting Speed, feet/min
$1/n$ = Cutting Speed Exponent
f = Feed Rate, in/rev
$1/m$ = Feed Rate Exponent
C = Taylor's Modified Tool Life Constant (min).

The cutting time can be expressed as:

$$t_c = B f^{-1} V^{-1} \tag{16.3}$$

where

B = Cutting Path Surface Factor (an input value), ft
f = Feed Rate, in/rev
V = Cutting Speed, feet/min.

The number of tool changes per piece can be found by:

$$n_t = Q t_c / T \tag{16.4}$$

where

Q = Fraction of cutting time that the tool is worn
t_c = Cutting Time, min/piece
T = Tool Life, min.

The value of Q is approximately 1.0 for operations such as turning, but for other metal cutting operations it may be as low as 0.10, say for horizontal milling operations.

Utilizing Equations (16.2) and (16.3) in Equation (16.4), one obtains:

$$n_t = Q B C^{-1} f^{(1/m-1)} V^{(1/n-1)} . \tag{16.5}$$

The total cost can be expressed as the sum of the cost components as:

$$\begin{aligned} C_u &= \text{Machine Cost} + \text{Operator Cost} + \text{Tool Cost} + \text{Tool Changing Cost} \\ &= R_m(t_c + t_l) + R_o(t_c + t_l) + C_t n_t + (R_o + R_m) t_{ch} n_t . \end{aligned} \tag{16.6}$$

After substituting for n_t, t_c, and T to obtain the cost expression in terms of V and f, one obtains:

$$\begin{aligned} C_u &= (R_o + R_m) t_l + (R_o + R_m) B f^{-1} V^{-1} \\ &\quad + [(R_o + R_m) t_{ch} + C_t] Q B C^{-1} f^{(1/m-1)} V^{(1/n-1)} . \end{aligned} \tag{16.7}$$

The constants can be combined and the unit cost objective function can be restated as:

$$C_u = K_{oo} + K_{01} f^{-1} V^{-1} + K_{02} f^{(1/m-1)} V^{(1/n-1)} \tag{16.7}$$

where

$$\begin{aligned} K_{oo} &= (R_o + R_m) t_l \\ K_{01} &= (R_o + R_m) B \\ K_{02} &= [(R_o + R_m) t_{ch} + C_t] Q B C^- . \end{aligned}$$

Since K_{oo} is a constant, the objective function to be minimized is:

$$Y = K_{01} f^{-1} V^{-1} + K_{02} f^{(1/m-1)} V^{(1/n-1)} \qquad (16.8)$$

where

Y = Variable portion of the unit cost.

The two constraints must be developed into geometric programming format. The feed constraint is:

$$f \leq f_{max} \qquad (16.9)$$

where

f_{max} = Maximum Feed Limit (in/rev).

The feed constraint is typically used to control the surface finish as the smaller the feed, the better the surface finish.

The horsepower constraint is given as:

$$a V^b f^c \leq Hp \qquad (16.10)$$

where

    a = Horsepower Constraint Constant
    b = Velocity Exponent for Horsepower Constraint
    c = Feed Rate Exponent for Horsepower Constraint
    HP = Horsepower Limit.

These constraints can be put into geometric programming form as:

$$K_{11} f \leq 1 \qquad (16.11)$$
$$K_{21} V^b f^c \leq 1 \qquad (16.12)$$

where

$$K_{11} = 1/f_{max}$$
$$K_{21} = a/Hp.$$

16.3 PROBLEM SOLUTION

The primal problem can be stated as:

$$\text{Minimize } Y = K_{01} f^{-1} V^{-1} + K_{02} f^{(1/m-1)} V^{(1/n-1)}. \qquad (16.13)$$

Subject to the constraints:

$$K_{11} f \leq 1 \qquad (16.14)$$
$$K_{21} V^b f^c \leq 1. \qquad (16.15)$$

16. MATERIAL REMOVAL/METAL CUTTING ECONOMICS

From the coefficients and signs, the signum values for the dual are:

$$\sigma_{01} = 1$$
$$\sigma_{02} = 1$$
$$\sigma_{11} = 1$$
$$\sigma_{21} = 1$$
$$\sigma_1 = 1$$
$$\sigma_2 = 1.$$

The dual problem can be formulated as:

$$\omega_{01} + \omega_{02} = 1 \qquad (16.16)$$
$$f \text{ terms} \quad -\omega_{01} + (1/m - 1)\omega_{02} + \omega_{11} + c\,\omega_{21} = 0 \qquad (16.17)$$
$$V \text{ terms} \quad -\omega_{01} + (1/n - 1)\omega_{02} + b\,\omega_{21} = 0. \qquad (16.18)$$

The degrees of difficulty (D) are equal to:

$$D = T - (N+1) = 4 - (2+1) = 1. \qquad (16.19)$$

From the constraint equations which have only one term it is apparent that:

$$\omega_{10} = \omega_{11} \qquad (16.20)$$
$$\omega_{20} = \omega_{21}. \qquad (16.21)$$

The dual objective function is:

$$Y = (K_{01}/\omega_{01})^{\omega_{01}} (K_{02}/\omega_{02})^{\omega_{02}} (K_{11})^{\omega_{11}} (K_{21})^{\omega_{21}}. \qquad (16.22)$$

Thus, we have 4 variables and 3 equations, or one degree of difficulty. To find an additional equation one must get the number of variables reduced to one in the dual objective function, differentiate the logarithm of the equation to obtain another equation by setting the derivative to zero. It was decided to determine the dual variables in terms of ω_{02}.

From Equation (16.16) one obtains:

$$\omega_{01} = 1 - \omega_{02}. \qquad (16.23)$$

From Equation (16.18) one obtains:

$$b\omega_{21} = \omega_{01} - (1/n - 1)\omega_{02}$$
$$= (1 - \omega_{02}) - (1/n - 1)\omega_{02}$$
$$= 1 - \omega_{02}/n$$
$$\omega_{21} = 1/b - \omega_{02}/bn. \qquad (16.24)$$

From Equation (16.17) one obtains:

$$\omega_{11} = \omega_{01} - (1/m - 1)\omega_{02} - c\omega_{21}$$
$$= 1 - \omega_{02} - (1/m - 1)\omega_{02} - c(1/b - \omega_{02}/bn)$$
$$= (1 - c/b) + \omega_{02}(c/bn - 1/m)$$
$$\omega_{11} = (1 - c/b) + \omega_{02} Z \tag{16.25}$$

where

$$Z = (c/bn - 1/m). \tag{16.26}$$

The dual can be written as:

$$Y = (K_{01}/(1-\omega_{02}))^{(1-\omega_{02})}(K_{02}/\omega_{02})^{\omega_{02}}(K_{11})^{((1-c/b)+\omega_{02}Z)}(K_{21})^{(1/b-\omega_{02}/bn)}. \tag{16.27}$$

The log of the dual is:

$$\text{Log}(Y) = (1-\omega_{02})\log(K_{01}/(1-\omega_{02})) + \omega_{02}\log(K_{02}/\omega_{02})$$
$$+ (1-c/b)\log(K_{11}) + \omega_{02}Z\log(K_{11})$$
$$+ 1/b\log(K_{21}) - \omega_{02}/bn\log(K_{21}). \tag{16.28}$$

Find $\partial(\log Y)/\partial\omega_{02}$ and set it to zero to find an additional equation.

$$\partial(\log Y)/\partial\omega_{02} = (1-\omega_{02})[1/\{K_{01}/(1-\omega_{02})\}][K_{01}(-1)(1-\omega_{02})^{-2}(-1)]$$
$$+ [\log(K_{01}/(1-\omega_{02}))](-1)$$
$$+ \omega_{02}[1/\{K_{02}/\omega_{02}\}][K_{02}(-1)\omega_{02}^{-2}(1)] + [\log(K_{02}/\omega_{02})](1)$$
$$+ 0 + Z\log(K_{11}) + 0 - (1/bn)\log(K_{21})$$
$$\partial(\log Y)/\partial\omega_{02} = (-1) - [\log(K_{01}/(1-\omega_{02}))] + 1 + [\log(K_{02}/\omega_{02})]$$
$$+ 0 + Z\log(K_{11}) + 0 - (1/bn)\log(K_{21})$$
$$\partial(\log Y)/\partial\omega_{02} = \log[(K_{02}/K_{01})((1-\omega_{02})/(\omega_{02}))] + \log(K_{11})^Z - \log(K_{21})^{1/bn}$$
$$\partial(\log Y)/\partial\omega_{02} = \log[(K_{02}/K_{01})((1-\omega_{02})/(\omega_{02}))] + \log[(K_{11})^Z/\log(K_{21})^{1/bn}]$$
$$\partial(\log Y)/\partial\omega_{02} = \log[(K_{02}/K_{01})((1-\omega_{02})/(\omega_{02}))(K_{11})^Z/(K_{21})^{1/bn}] \tag{16.29}$$

Setting the expression equal to zero and then use the anti-log, one goes through various steps to obtain an expression for ω_{02}:

$$1 = (K_{02}/K_{01})((1-\omega_{02})/(\omega_{02}))(K_{11})^Z/(K_{21})^{1/bn}$$
$$(1-\omega_{02})/(\omega_{02}) = (K_{01}/K_{02})((K_{21})^{1/bn}/(K_{11})^Z)$$
$$1/\omega_{02} - 1 = (K_{01}/K_{02})((K_{21})^{1/bn}/(K_{11})^Z)$$
$$1/\omega_{02} = (K_{01}/K_{02})((K_{21})^{1/bn}/(K_{11})^Z) + 1$$
$$\omega_{02} = (K_{02}K_{11}^Z)/(K_{01}K_{21}^{1/bn} + K_{02}K_{11}^Z). \tag{16.30}$$

Now the values for other three dual variables can be obtained in terms of ω_{02} as:

$$\omega_{01} = -\omega_{02} \tag{16.31}$$
$$\omega_{21} = 1/b - \omega_{02}/bn \tag{16.32}$$
$$\omega_{11} = (1 - c/b) + \omega_{02}Z. \tag{16.33}$$

16. MATERIAL REMOVAL/METAL CUTTING ECONOMICS

Substituting the values of the dual variables into the dual objective function, the dual objective function can be expressed in terms of ω_{02} and is:

$$Y = (K_{01}/(1 - \omega_{02}))^{(1-\omega_{02})}(K_{02}/\omega_{02})^{\omega_{02}}(K_{11})^{((1-c/b)+\omega_{02}Z)}(K_{21})^{(1/b-\omega_{02}/bn)} \qquad (16.34)$$

and the minimum cost would be:

$$C_u = Y + K_{00} \qquad (16.35)$$

$$\text{or} \quad C_u = K_{00} + (K_{01}/(1 - \omega_{02}))^{(1-\omega_{02})}(K_{02}/\omega_{02})^{\omega_{02}}$$
$$(K_{11})^{((1-c/b)+\omega_{02}Z)}(K_{21})^{(1/b-\omega_{02}/bn)}. \qquad (16.36)$$

By using the primal-dual relationships of the objective function, one can obtain the following equation for the primal variables. Starting with

$$K_{01} f^{-1} V^{-1} = \omega_{01} Y \qquad (16.37)$$

and

$$K_{02} f^{1/m-1} V^{1/n-1} = \omega_{02} Y. \qquad (16.38)$$

One can obtain equations for the primal variables in terms of the dual variables, dual objective function and constants. These equations are for the case when both constraints are binding.

$$f = (\omega_{02} K_{01}/(\omega_{01} K_{02}))^m (K_{01}/\omega_{01})^{(m(m-1)/(n-m))}$$
$$(\omega_{02} K_{02})^{((mm)/(n-m))} Y^{(m/(n-m))} \qquad (16.39)$$
$$V = (K_{01}/\omega_{01})^{(n(m-1)/(m-n))} (\omega_{02}/K_{02})^{((mn)/(m-n))} Y^{(n/(m-n))}. \qquad (16.40)$$

Upon examination of the equations for the dual variables, it is possible for either ω_{11} or ω_{21} to be negative, which means that the constraints are not binding. Thus, the dual variable must be set to zero if the constraint is not binding. If ω_{11} is zero, then Equations (16.16), (16.17), and (16.18) become:

	$\omega_{01} +$	ω_{02}	$= 1$	(16.41)
f terms	$-\omega_{01} + (1/m - 1)\omega_{02}$		$+ c\,\omega_{21} = 0$	(16.42)
V terms	$-\omega_{01} + (1/n - 1)\,\omega_{02}$		$+ b\,\omega_{21} = 0.$	(16.43)

The degree of difficulty becomes zero with the variable ω_{11} removed, and the new values for the dual variables become:

$$\omega_{02} = (b - c)/(b/m - c/n) \qquad (16.44)$$
$$\omega_{01} = 1 - \omega_{02} \qquad (16.45)$$
$$\omega_{11} = 0 \quad \text{(loose constraint)} \qquad (16.46)$$
$$\omega_{21} = (1/m - 1/n)/(b/m - c/n). \qquad (16.47)$$

If ω_{21} is zero, then Equations (16.16), (16.17), and (16.18) become:

$$\omega_{01} + \omega_{02} = 1 \quad (16.48)$$
$$f \text{ terms} \quad -\omega_{01} + (1/m - 1)\omega_{02} + \omega_{11} = 0 \quad (16.49)$$
$$V \text{ terms} \quad -\omega_{01} + (1/n - 1)\omega_{02} = 0. \quad (16.50)$$

The degree of difficulty becomes zero with the variable ω_{21} removed, and the new values for the dual variables become:

$$\omega_{02} = n \quad (16.51)$$
$$\omega_{01} = 1 - \omega_{02} \quad (16.52)$$
$$\omega_{11} = 1 - n/m \quad (16.53)$$
$$\omega_{21} = 0. \quad (16.54)$$

The equations for velocity and feed for this set of dual variables are:

$$V = (n/(1-n))^n (K_{01}/K_{02})^n K_{11}^{(n/m)} \quad (16.55)$$
$$f = (K_{01}/(\omega_{01} Y V)) \quad (16.56)$$

Thus, there are three different solutions based upon the values of the constants in the problem. However, the equations for the dual variables for the three possibilities have been obtained and the only question is to determine which of the three conditional are applicable. The results for the primal variables have been given for the case when both constraints are binding and when the feed constraint is binding (horsepower is non-binding). The horsepower constraint is often not binding, but the feed constraint is usually a binding constraint.

16.4 EXAMPLE PROBLEM

An example problem will be used to illustrate the application of the formulas and the magnitude of the results obtained. The data used is from Ermer (2) and results obtained are in Table 16.1.

The model results indicate that the cutting speed is 290 ft/min and the feed rate is at the max of 0.005 in/rev. The variable cost (Y) is $1.05 and the total cost is $1.25.

16.5 EVALUATIVE QUESTIONS

1. In a problem with two constraints, how many solutions are possible and describe the conditions under which the solutions would occur?

2. The data for the metal cutting problem has been modified and is in Table 16.2. Calculate the values for the constants and constants and variables.

3. The data for the metal cutting problem has been modified and is in Table 16.3. Calculate the values for the constants and constants and variables. However, for this problem consider the horsepower to be non-binding.

Table 16.1: Data for Metal Cutting Example Problem

Input Parameters		Calculated Constants and Variables	
Q	1	K_{00}	0.20
R_0	0.04 \$/min	K_{01}	1.2566
R_m	0.06 \$/min	K_{02}	$1.80 * 10^{-8}$
C_t	0.50 \$/edge	K_{11}	200
t_h	2 min	K_{21}	0.358
t_{ch}	0.50 min	Z	2.2684
D	6 inches	ω_{01}	0.8216
L	8 inches	ω_{02}	0.1784
m	0.862	ω_{11}	0.5476
n	0.25	ω_{21}	0.3146
C	$3.84 * 10^8$ min	Y	\$ 1.05
a	3.58	C_u	\$ 1.25
b	0.91	B	12.56 in-ft $= \pi D$ (in) L (ft)
c	0.78	f	0.005 in/rev
Hp	10	V	290 ft/min
f_{max}	0.005 in/rev		

REFERENCES

[1] Robert C. Creese and Pingfang Tsai, "Generalized Solution for Constrained Metal Cutting Economics Problem", *1985 Annual International Industrial Conference Proceedings*, Institute of Industrial Engineers U.S.A, 113–117. 85

[2] Ermer, D.S., "Optimization of the Constrained Machining Economics Problem by Geometric Programming," *Journal of Engineering for Industry*, Traisactions of the ASME, November 1971, pp 1067–1072. 85

[3] Tsai, Pingfang *An Optimization Algorithm and Economic Analysis for a Constrained Machining Model*, PhD Dissertation, West Virginia University, Morgantown, WV, 214pp. 85

Table 16.2: Data for Problem 2

Input Parameters		Calculated Constants and Variables		
Q	1	K_{00}		
R_0	0.4 \$/min	K_{01}		
R_m	0.6 \$/min	K_{02}		
C_t	2.0 \$/edge	K_{11}	200	
t_h	1.5 min	K_{21}	0.358	
t_{ch}	0.80 min	Z		
D	1 inches	ω_{01}		
L	6 inches	ω_{02}		
m	0.80	ω_{11}		
n	0.25	ω_{21}		
C	$5.00 * 10^8$ min	Y	1.14	
a	3.58	C_u		
b	0.91	B		
c	0.78	f	0.005 in/rev	
Hp	10	V	290 ft/min	
f_{max}	0.005			

Table 16.3: Data for Problem 3 (horsepower constraint removed)

Input Parameters		Calculated Constants and Variables		
Q	1	K_{00}		
R_0	0.4 \$/min	K_{01}		
R_m	0.6 \$/min	K_{02}		
C_t	2.0 \$	K_{11}	200	
t_h	1.5 min	K_{21}	0.358	
t_{ch}	0.80 min	Z		
D	1 inches	ω_{01}		
L	6 inches	ω_{02}		
m	0.80	ω_{11}		
n	0.25	ω_{21}		
C	$5.00 * 10^8$ min	Y	0.91	
a	3.58	C_u		
b	0.91	B		
c	0.78	f	0.005 in/rev	
Hp	NA	V	460 ft/min	
f_{max}	0.005			

CHAPTER 17

The Open Cargo Shipping Box with Skids

17.1 INTRODUCTION

The open cargo shipping box was presented in Chapter 6 and is the classical problem in geometric programming. It had zero degrees of difficulty and was solved relatively easily. The open cargo shipping box problem is adjusted to add skid rails at $5/unit length and if two rails are used, the additional cost would be $10/box length. This becomes a problem with one degree of difficulty and the solution is more difficult, and the purpose of this chapter is to introduce different methods of obtaining a solution.

The first method to be applied is the *constrained derivative approach*, that is obtaining four of the dual variables in terms of the fifth variable, taking the derivative of the function and setting it to zero and obtaining a fifth independent dual equation. The problem then becomes a problem with zero degrees of difficulty and the solution is obtained.

The second method applied is the *dimensional analysis approach*, which takes the primal dual equations and separating them into the primal variables on one side of the equation and the dual variables and constants on the other side. Solving this for the exponents of the terms leads to the required additional independent equation. The second method does not require the taking of the derivatives of logarithmic functions which can be difficult. In some manners, the approach is similar to the method of solving the dual equations. One advantage of the dimensional analysis method is that students are often familiar with this technique in evaluating units engineering problems.

The third method is the *condensation of terms approach*. This method combines two terms to reduce the degrees of difficulty by one. This method, however, does not guarantee an optimal solution and the selection of which terms to combine is important. The terms selected should have exponents that close in value and should be selected such that the equations in the new dual do not result in dual variables becoming zero or that redundant equations result in the formation of the dual. The terms are typically combined geometrically with equal weights.

17.2 PRIMAL-DUAL PROBLEM FORMULATION

The problem can be stated as: "400 cubic yards of gravel must be ferried across a shallow river. The box is an open box with skids used because of the low water level. The box has length L, width W, and depth H. The sides of the box cost $10/yd² and have a total area of $2(L + W)H$ and the bottom

17. THE OPEN CARGO SHIPPING BOX WITH SKIDS

of the box cost \$20/yd² and has an area of LW. There are 2 skids at a cost of \$5/yd on the length of the box and the total skid length is $2L$. Each round trip of the box on the ferry will cost 10 cents per cubic yard of gravel shipped." The objective function becomes:

$$C = 40/LWH + 10LW + 20LH + 40HW + 10L . \qquad (17.1)$$

The previous cost solution was \$100 and the box length was 2 units (yds), so the new cost would be \$100 + \$20 = \$120 if the same dimensions are used. However, one must determine if the additional cost yields the same dimensions ($L = 2$ yds, $W = 1$ yd, $H = 1/2$ yd) and 400 trips were required.

The primal objective function can be written in general terms of:

$$C = C_{01}/(LWH) + C_{02}LW + C_{03}LH + C_{04}HW + C_{05}L . \qquad (17.2)$$

The degrees of difficulty is

$$5 - (3 + 1) = 1 . \qquad (17.3)$$

The dual formulation is:

$$\begin{array}{lll} \text{Objective Function} & \omega_{01} + \omega_{02} + \omega_{03} + \omega_{04} + \omega_{05} = 1 & (17.4) \\ L \text{ terms} & -\omega_{01} + \omega_{02} + \omega_{03} \quad\quad + \omega_{05} = 0 & (17.5) \\ W \text{ terms} & -\omega_{01} + \omega_{02} \quad\quad + \omega_{04} \quad = 0 & (17.6) \\ H \text{ terms} & -\omega_{01} \quad\quad + \omega_{03} + \omega_{04} \quad = 0 . & (17.7) \end{array}$$

Subtracting Equation (17.7) from Equation (17.6) one obtains that:

$$\omega_{02} = \omega_{03} . \qquad (17.8)$$

Considering Equation (17.4) and Equation (17.5), one obtains

$$\omega_{04} = 1 - 2\omega_{01} . \qquad (17.9)$$

Using Equation (17.6) with Equations (17.7) and (17.8), one obtains

$$\omega_{02} = 3\omega_{01} - 1 \qquad (17.10)$$

and using Equation (17.4) with Equations (17.8) thru (17.10), one obtains

$$\omega_{05} = 2 - 5\omega_{01} . \qquad (17.11)$$

The dual can be written as:

$$\text{Dual}(Y) = \{(C_{01}/\omega_{01})^{\omega_{01}}(C_{02}/\omega_{02})^{\omega_{02}}(C_{03}/\omega_{03})^{\omega_{03}}(C_{04}/\omega_{04})^{\omega_{04}}(C_{05}/\omega_{05})^{\omega_{05}}\} \qquad (17.12)$$

$$Y = \{(C_{01}/\omega_{01})^{\omega_{01}}(C_{02}/(3\omega_{01}-1))^{(3\omega_{01}-1)}(C_{03}/(3\omega_{01}-1))^{(3\omega_{01}-1)}$$
$$(C_{04}/(1-2\omega_{01}))^{(1-2\omega_{01})}(C_{05}/(2-5\omega_{01}))^{(2-5\omega_{01})}\} . \qquad (17.13)$$

17.3 CONSTRAINED DERIVATIVE APPROACH

The constrained derivate approach takes the derivative of the log Y with respect to ω_{01}, sets it to zero and solve for ω_{01}.

$$\text{Log } Y = \omega_{01} \log(C_{01}/\omega_{01}) + (3\omega_{01} - 1)\log(C_{02}/(3\omega_{01} - 1)) + (3\omega_{01} - 1)$$
$$\log(C_{03}/(3\omega_{01} - 1)) + (1 - 2\omega_{01})\log(C_{04}/(1 - 2\omega_{01}))$$
$$+ (2 - 5\omega_{01})\log(C_{05}/(2 - 5\omega_{01})) \quad (17.14)$$

$$\partial(\log Y)/\partial(\omega_{01}) = 0 = -1 + \log(C_{01}/\omega_{01}) + -3 + 3\log(C_{02}/(3\omega_{01} - 1)) - 3$$
$$+ 3\log(C_{03}/(3\omega_{01} - 1)) + 2 - 2\log(C_{04}/(1 - 2\omega_{01}))$$
$$+ 5 - 5\log(C_{05}/(2 - 5\omega_{01})) \quad (17.15)$$

$$= \log(C_{01}/\omega_{01}) + 3\log(C_{02}/(3\omega_{01} - 1)) + 3\log(C_{03}/(3\omega_{01} - 1))$$
$$- 2\log(C_{04}/(1 - 2\omega_{01})) - 5\log(C_{05}/(2 - 5\omega_{01})) \quad (17.16)$$

The antilog is taken and the constants are put on one side and the dual variable equations on the other to obtain:

$$C_{01}C_{02}^3 C_{03}^3/(C_{04}^2 C_{05}^5) = Z = \omega_{01}(3\omega_{01} - 1)^3(3\omega_{01} - 1)^3/(1 - 2\omega_{01})^2(2 - 5\omega_{01})^5 \quad (17.17)$$
$$Z = 40 \times 10^3 \times 20^3/(40^2 \times 10^5) = 2 \quad (17.18)$$
$$\omega_{01}(3\omega_{01} - 1)^3(3\omega_{01} - 1)^3/(1 - 2\omega_{01})^2(2 - 5\omega_{01})^5 = 2. \quad (17.19)$$

Note that the terms of Equation (17.19) must be positive, so that indicates that ω_{01} must be $< 1, > 1/3, < 0.5$, and < 0.4 which indicates it is between 1/3 and 0.4.

If one solves Equation (17.19) for ω_{01} (using search techniques), the value obtained was:

$$\omega_{01} = 0.3776519. \quad (17.20)$$

Using Equations (17.8) to (17.11) for the other variables in terms of ω_{01}; one obtains

$$\omega_{01} = 0.378 \quad (17.21)$$
$$\omega_{02} = 0.133 \quad (17.22)$$
$$\omega_{03} = 0.133 \quad (17.23)$$
$$\omega_{04} = 0.245 \quad (17.24)$$
$$\omega_{05} = 0.111. \quad (17.25)$$

Using the values of C and ω in the Equation (17.12) for the dual, one obtains

$$Y = (40/0.378)^{0.378}(10/0.133)^{0.133}(20/0.133)^{0.133}(40/0.245)^{0.245}$$
$$(10/0.111)^{0.111} = \$115.72. \quad (17.26)$$

This compares with the cost of $100 for the initial problem without skids.
Using the primal-dual relationships and the objective function, the primal variables can be found as:

$$L = \omega_{05}Y/10 = (.111)(115.72)/10 = 1.284 \text{ yd (versus 2) or } \omega_{05}Y/C_{05} \quad (17.27)$$

98 17. THE OPEN CARGO SHIPPING BOX WITH SKIDS

Similarly,

$$H = \omega_{03}Y/(20L) = \omega_{03}Y/(20\omega_{05}Y/10) = (\omega_{03}/\omega_{05})(C_{05}/C_{03})$$
$$= (0.133/0.111)(10/20) = 0.599 \quad \text{(versus 0.500)} \tag{17.28}$$

and

$$W = \omega_{02}Y/(10L) = \omega_{02}Y/(10\omega_{05}Y/10) = (\omega_{02}/\omega_{05})(C_{05}/C_{02})$$
$$= (0.133/0.111)(10/10) = 1.198 \quad \text{(versus 1.000)} \tag{17.29}$$

Therefore, the primal becomes

$$C = C_{01}/(LWH) + C_{02}LW + C_{03}LH + C_{04}HW + C_{05}L \tag{17.30}$$
$$= C_{01}/(1.284*0.599*1.198) + C_{02}(1.284*1.198) + C_{03}(1.284*0.599)$$
$$+ C_{04}(0.599*1.198) + C_{05}*(1.284)$$
$$= 43.41 + 15.38 + 15.38 + 28.70 + 12.84$$
$$= 115.71 \text{ versus } 115.72. \tag{17.31}$$

The volume of the box is $1.284 \times 0.599 \times 1.198 = 0.921$ cubic yards versus the 1.0 cubic yard volume in the original problem. The number of trips will increase from 400 in the original problem to 434.3 or 435 trips. The shipping cost increases from \$40 to \$43.41, but the box cost (side, end and bottom) totals \$59.46 versus \$60 and the primary increase is the cost of the skids, that is \$12.84.

17.4 DIMENSIONAL ANALYSIS APPROACH FOR ADDITIONAL EQUATION

The dimensional analysis approach starts with the primal and dual equations which are:

$$C = C_{01}/(LWH) + C_{02}LW + C_{03}LH + C_{04}HW + C_{05}L \tag{17.2}$$

and

$$\text{Dual }(Y) = \{(C_{01}/\omega_{01})^{\omega_{01}}(C_{02}/\omega_{02})^{\omega_{02}}(C_{03}/\omega_{03})^{\omega_{03}}(C_{04}/\omega_{04})^{\omega_{04}}(C_{05}/\omega_{05})^{\omega_{05}}\} \tag{17.12}$$

The primal dual relationships are:

$$L^{-1}W^{-1}H^{-1} = \omega_{01}Y/C_{01} \tag{17.32}$$
$$LW = \omega_{02}Y/C_{02} \tag{17.33}$$
$$LH = \omega_{03}Y/C_{03} \tag{17.34}$$
$$HW = \omega_{04}Y/C_{04} \tag{17.35}$$
$$L = \omega_{03}Y/C_{03} \tag{17.36}$$

These relationships can be combined to give:

$$(L^{-1}W^{-1}H^{-1})^A(LW)^B(LH)^C(HW)^D(L)^E = 1$$
$$= (\omega_{01}Y/C_{01})^A(\omega_{02}Y/C_{02})^B(\omega_{03}Y/C_{03})^C(\omega_{04}Y/C_{04})^D(\omega_{05}Y/C_{05})^E \tag{17.37}$$

17.4. DIMENSIONAL ANALYSIS APPROACH FOR ADDITIONAL EQUATION

The equations from dimensional analysis are:

$$L \text{ terms} \quad -A + B + C \quad\quad + E = 0 \tag{17.38}$$
$$W \text{ terms} \quad -A + B \quad\quad + D \quad\quad = 0 \tag{17.39}$$
$$H \text{ terms} \quad -A \quad\quad + C + D \quad\quad = 0 \tag{17.40}$$
$$Y \quad\quad +A + B + C + D + E = 0. \tag{17.41}$$

From Equations (17.39) and (17.40), one obtains that:

$$B = C \tag{17.42}$$

Subtracting Equation (17.38) from Equation (17.41), one obtains:

$$D = -2A \tag{17.43}$$

Subtracting Equation (17.38) from Equation (17.39) results in

$$D = C + E \tag{17.44}$$

From Equation (17.38) one observes that:

$$C + E = A - B \tag{17.45}$$

Thus,

$$D = -2A = A - B, \tag{17.46}$$

so

$$B = 3A \tag{17.47}$$
$$C = 3A \tag{17.48}$$
$$D = -2A \tag{17.49}$$

and via Equation (17.38)

$$E = -5A. \tag{17.50}$$

Thus, if $A = 1$, then $B = 3$, $C = 3$, $D = -2$, and $E = -5$.
Thus, from Equation (17.37)

$$1 = (\omega_{01} Y / C_{01})^1 (\omega_{02} Y / C_{02})^3 (\omega_{03} Y / C_{03})^3 (\omega_{04} Y / C_{04})^{-2} (\omega_{05} Y / C_{05})^{-5}. \tag{17.51}$$

This results in

$$(\omega_{01})(\omega_{02})^3 (\omega_{03})^3 (\omega_{01})^{-2} (\omega_{01})^{-5} = (C_{01})(C_{02})^3 (C_{03})^3 (C_{04})^{-2} (C_{05})^{-5}$$
$$= (40)(2)^3 (10)^3 (40)^{-1} (10)^{-5} \tag{17.52}$$

or in terms of ω_{01}

$$(\omega_{01})(3\omega_{01} - 1)^3 (3\omega_{01} - 1)^3 (1 - 2\omega_{01})^{-2} (2 - 5\omega_{01})^5 = 2. \tag{17.53}$$

Equation (17.53) is the same as Equation (17.19) and the solution procedure from that point on would be the same as that used for the constrained derivative approach.

17.5 CONDENSATION OF TERMS APPROACH

The technique of condensation involves the combining of terms to reduce the degrees of difficulty to make the solution easier. D.J. Wilde presented this technique in his book "Globally Optimum Design" in 1978 [1] and this example is presented in "Engineering Design – A Material and Processing Approach" by George Dieter (1991) [2].

The primal problem was:

$$C = 40/LWH + 10LW + 20LH + 40HW + 10L \qquad (17.1)$$

or as

$$C = C_{01}/(LWH) + C_{02}LW + C_{03}LH + C_{04}HW + C_{05}L \qquad (17.2)$$

The dual is

$$\text{Dual}(Y) = \{(C_{01}/\omega_{01})^{\omega_{01}} (C_{02}/\omega_{02})^{\omega_{02}} (C_{03}/\omega_{03})^{\omega_{03}} (C_{04}/\omega_{04})^{\omega_{04}} (C_{05}/\omega_{05})^{\omega_{05}}\}, \qquad (17.12)$$

and there is one degree of difficulty.

The condensation technique combines two terms to reduce the degrees of difficulty. The selection of the terms and the weighing of the terms is important. The terms selected should have exponents that are not that different and should be selected such that equations in the new dual do not result in dual variables resulting to be zero or that redundant equations result in the formulation of the dual. The weights for combining the two terms will be considered to be equal. The formation of the new term will be illustrated.

If the third and fifth terms are combined with equal weights, the result is

$$t_3 + t_5 \geq (C_{03}LH/(1/2))^{1/2}(C_{05}L/(1/2))^{1/2} = 2(C_{03}C_{05})^{1/2}LH^{1/2} = C_{06}LH^{1/2} \qquad (17.53)$$

where

$$C_{06} = 2(C_{03}C_{05})^{1/2}. \qquad (17.54)$$

The new primal would be:

$$C = C_{01}/(LWH) + C_{02}LW + C_{04}HW + C_{06}LH^{1/2}. \qquad (17.55)$$

The new Dual would be

$$\text{Dual}(Y) = \{(C_{01}/\omega_{01})^{\omega_{01}} (C_{02}/\omega_{02})^{\omega_{02}} (C_{04}/\omega_{04})^{\omega_{04}} (C_{06}/\omega_{06})^{\omega_{06}}\}. \qquad (17.56)$$

The new dual formulation is:

Objective Function	$\omega_{01} + \omega_{02} + \omega_{04} + \omega_{06} = 1$	(17.57)
L terms	$-\omega_{01} + \omega_{02} + \omega_{06} = 0$	(17.58)
W terms	$-\omega_{01} + \omega_{02} + \omega_{04} = 0$	(17.59)
H terms	$-\omega_{01} + \omega_{04} + \omega_{06}/2 = 0$.	(17.60)

Equations (17.58) and (17.59) result in:

$$\omega_{06} = \omega_{04} \tag{17.61}$$

Equation (17.60) and (17.61) result in:

$$\omega_{01} = \omega_{04} + \omega_{06}/2 = (3/2)\omega_{04} \tag{17.62}$$

Using Equations (17.61) and (17.62) in Equation (17.58) results in:

$$\omega_{02} = \omega_{01} - \omega_{06} = \omega_{04}/2 \tag{17.63}$$

Using Equations (17.61), (17.62), and (17.63) with Equation (17.59) results in:

$$\omega_{04} = 0.25 \tag{17.64}$$
$$\omega_{06} = 0.25 \tag{17.65}$$
$$\omega_{01} = 0.375 \tag{17.66}$$
$$\omega_{02} = 0.125 . \tag{17.67}$$

Using the values for the constants and the dual variables, the dual objective function becomes:

$$\text{Dual}(Y) = \{(40/0.375)^{0.375}(10/0.125)^{0.125}(40/0.25)^{0.25}(2(20\times 10)^{1/2}/0.25)^{0.25}\}$$
$$= (5.761)(1.729)(3.557)(3.261) = 115.5 \text{ (versus } 115.71). \tag{17.68}$$

The primal dual relationships yield

$$C_{01}/HWL = \omega_{01}Y \tag{17.69}$$
$$C_{02}LW = \omega_{02}Y \tag{17.70}$$
$$C_{04}HW = \omega_{04}Y \tag{17.71}$$
$$C_{06}LH^{1/2} = \omega_{06}Y . \tag{17.72}$$

Combining Equations (17.70) and (17.71) result in:

$$H/L = (C_{02}/C_{04})(\omega_{04}/\omega_{02}) = (10/40)(0.25/0.125) = 1/2 . \tag{17.73}$$

Combining Equations (17.70) and (17.72) result in:

$$(W/H^{1/2}) = (C_{06}/C_{02})(\omega_{02}/\omega_{06})$$
$$= (2(20\times 10)^{1/2}/(10)\times(0.125/0.25)) = 2^{1/2} = 1.414 . \tag{17.74}$$

Combining Equations (17.69) and (17.70) result in

$$L^2W^2H = (C_{01}/C_{02})(\omega_{02}/\omega_{01}) = (40/10)(0.125/0.375) = 4/3 \tag{17.75}$$

Using Equation (17.73) and 17.74 with Equation (17.75) one obtains:

$$(2H)^2(2H)H = 4/3 \tag{17.76}$$

and this results in:

$$H = (1/6)^{0.25} = 0.6389 \text{ yd (vs } 0.599 \text{ yd in constrained derivative solution)} \quad (17.77)$$
$$L = 2H = 1.2779 \text{ yd (vs } 1.284 \text{ yd in constrained derivative solution)} \quad (17.78)$$
$$W = (2H)^{1/2} = 1.1304 \text{ yd (vs } 1.198 \text{ yd in constrained derivative solution)} \quad (17.79)$$

The primal objective function would be:

$$\begin{aligned} C &= C_{01}/(LWH) + C_{02}LW + C_{04}HW + C_{06}LH^{1/2} \quad (17.80) \\ &= 40/(1.2779 \times 1.1304 \times 0.6389) + 10(1.2779 \times 1.1304) + 40(0.6389 \times 1.1304) \\ &\quad + 20 \times 2^{1/2}(1.2779 \times 0.6389^{1/2}) \\ &= 43.34 + 14.45 + 28.89 + 28.89 \\ &= 115.57 \, . \quad (17.81) \end{aligned}$$

The values for the primal and dual are in good agreement and in good agreement with the constrained derivative approach and the dimensional analysis approach. The primal variables for the condensed version are slightly different than the primal variables of the other solutions. This is because the objective function is slightly different, but the values of the various objective functions are quite close.

17.6 EVALUATIVE QUESTIONS

1. If the cost of the skid rails was $20 instead of $10, what is the effect on the total cost, the number of trips, and the box dimensions?

2. Use the dimensional analysis approach on the metal cutting problem with two constraints to derive Equation (16.30).

3. Use the condensation technique with second and fifth terms of the primal objective function and compare the solutions of the original problem and when the condensation technique was used on terms 3 and 5.

REFERENCES

[1] D. J. Wilde, *Globally Optimum Design,* Wiley-Interscience, New York, 1978, pp. 88–90. 100

[2] George E. Dieter, *Engineering Design A Materials and Processing Approach*, 2nd Edition, McGraw-Hill, New York, 1991, pp. 223–225. 100

CHAPTER 18

Profit Maximization Considering Decreasing Cost Functions of Inventory Policy

18.1 INTRODUCTION

The classical inventory models consider the unit costs and unit prices to be fixed, but as the quantity increases the economies of scale permit decreased in costs and prices. In the model presented by Jung and Klein [1] in 2001, they considered cost per unit and price per unit to be power functions of demand. The variables in the profit maximization model considered are the order quantity and the product demand. The inventory model considered is the basic model with the assumptions that (1) replenishment is instantaneous; (2) no shortage is allowed; and (3) the order quantity is a batch.

18.2 MODEL FORMULATION

The variables and parameters used in the model are listed in Table 18.1. The variables are demand per unit time and order quantity. The parameter price is a function of the price scaling constant, the demand, and the price elasticity constant. The parameter cost is a function of the cost scaling constant, the demand, and the cost elasticity constant, also called the economy of scale factor.

The profit maximization model can be stated on a per unit time basis as:

$$\begin{aligned}\text{Maximize Profit } (\pi) = {} & \text{Total Variable Revenue } (R) \\ & - [\text{Total Variable Cost } (TVC) \\ & + \text{Total Set-up Cost } (TSC) \\ & + \text{Inventory Holding Cost } (IHC)]. \end{aligned} \quad (18.1)$$

The revenue per unit can be represented as:

$$P = aD^{-\alpha} \quad (18.2)$$

where

D = Demand per Unit Time
a = Scaling Constant for Price
α = Price Elasticity with respect to Demand .

18. PROFIT MAXIMIZATION

Table 18.1: Variables and parameters for inventory model

Symbol Used	Description of Variable or Parameter
D	Demand per Unit Time (decision variable)
Q	Order Quantity (decision variable)
P	Price per Unit ($/Unit)
C	Cost per Unit ($/Unit)
A	Set-up Cost ($/Batch)
i	Inventory Holding Cost Rate (Percent/Unit Time)
a	Scaling Constant for Price (Initial Price)
b	Scaling Constant for Cost (Initial Cost)
α	Price Elasticity with respect to Demand
β	Cost Elasticity with respect to Demand (Degree of Economies of Scale)
π	Profit per unit time
R	Total Revenue
TVC	Total Variable Cost
TSC	Total Set-up Cost

The total revenue (R) is the product of the demand and the revenue per unit:

$$R = D * P = D * aD^{-\alpha} = aD^{1-\alpha}. \tag{18.3}$$

The cost per unit (C) can be represented as:

$$C = bD^{-\beta} \tag{18.4}$$

where

$$D = \text{Demand per Unit Time}$$
$$b = \text{Scaling Constant for Cost}$$
$$\beta = \text{Cost Elasticity with respect to Demand}.$$

The total variable cost (TVC) is the product of the demand and the cost per unit:

$$TVC = D * C = D * bD^{-\beta} = bD^{1-\beta}. \tag{18.5}$$

The total set-up cost (TSC) is the product of the set-up cost times the number of set-ups which can be expressed as:

$$TSC = A * D/Q = ADQ^{-1}. \tag{18.6}$$

The inventory holding cost (IHC) represents the product of the average inventory ($Q/2$), the unit cost ($bD^{-\beta}$) and the inventory holding cost rate (i) and results in:

$$IHC = Q/2 * bD^{-\beta} * i = (ib/2) * QD^{-\beta}. \tag{18.7}$$

18.2. MODEL FORMULATION

Inserting Equations (18.3), (18.5), (18.6) and (18.7) into Equation (18.1), the result is the primal objective function:

$$\text{Max } (\pi) = aD^{1-\alpha} - (bD^{1-\beta} + ADQ^{-1} + (ib/2)QD^{-\beta}). \tag{18.8}$$

The problem can be written in general terms, such as

$$\text{Max } (\pi) = C_{00}D^{1-\alpha} - C_{01}D^{1-\beta} - C_{02}DQ^{-1} - C_{03}QD^{-\beta}. \tag{18.9}$$

To solve the problem one minimizes the negative of the profit function, that is, the primal objective function becomes:

$$\text{Min } Y = -C_{00}D^{1-\alpha} + C_{01}D^{1-\beta} + C_{02}DQ^{-1} + C_{03}QD^{-\beta} \tag{18.10}$$

where $Y = -\pi$.

The signum functions would be:

$$\sigma_{01} = -1$$
$$\sigma_{02} = \sigma_{03} = \sigma_{04} = 1,$$
and $\sigma_{00} = -1$ as this is a maximization problem.

The dual objective function is:

$$D(Y) = -1[(C_{00}/\omega_{01})^{-\omega_{01}}(C_{01}/\omega_{02})^{\omega_{02}}(C_{02}/\omega_{03})^{\omega_{03}}(C_{03}/\omega_{04})^{\omega_{04}}]^{-1}. \tag{18.11}$$

The dual formulation would be:

Objective Function $-\quad \omega_{01} + \quad \omega_{02} + \omega_{03} + \quad \omega_{04} = -1$	(18.12)
D Terms $\quad -(1-\alpha)\omega_{01} + (1-\beta)\omega_{02} + \omega_{03} - \beta\omega_{04} = 0$	(18.13)
Q Terms $\quad\quad\quad\quad\quad\quad\quad\quad\quad\quad -\omega_{03} + \quad \omega_{04} = 0.$	(18.14)

The degree of difficulty (D) is:

Degrees of Difficulty = Dual Variables(4) − (Primal Variables(2) + 1) = 1.

Since the degrees of difficult is 1, one additional equation is needed to solve the problem. First one must get the four dual variables in terms of one variable and then obtain an equation in terms of that variable. If one examines Equation (18.14) one observes that:

$$\omega_{03} = \omega_{04}. \tag{18.15}$$

Thus, one selects ω_{04} as the unknown variable and now must find ω_{01} and ω_{02} in terms of ω_{04}. If one multiplies Equation (18.12) by $(1-\alpha)$ and subtracts it from Equation (18.13) and using Equation (18.15), one obtains:

$$\omega_{02}[(1-\beta) - (1-\alpha)] + \omega_{04}[(1-\beta) - 2(1-\alpha)] = 1 - \alpha. \tag{18.16}$$

18. PROFIT MAXIMIZATION

This results in:

$$\omega_{02} = [(1-\alpha)/(\alpha-\beta)] + [(1+\beta-2\alpha)/(\alpha-\beta)]\omega_{04}. \tag{18.17}$$

Now if one solves for ω_{01} via Equations (18.12) and (18.15) one obtains:

$$\omega_{01} = 1 + \omega_{02} + 2\omega_{04}. \tag{18.18}$$

Now using Equation (18.17) in Equation (18.18), one obtains the expression for ω_{01} in terms of ω_{04} which is:

$$\omega_{01} = [(1-\beta)/(\alpha-\beta)][1+\omega_{04}]. \tag{18.19}$$

To obtain the additional equation, the approach of dimensional analysis will be used. The additional equations from the primal dual relationships are:

$$C_{00}D^{1-\alpha} = \omega_{01}Y \tag{18.20}$$
$$C_{01}D^{1-\beta} = \omega_{02}Y \tag{18.21}$$
$$C_{02}DQ^{-1} = \omega_{03}Y \tag{18.22}$$
$$C_{03}D^{-\beta}Q = \omega_{04}Y. \tag{18.23}$$

The dimensional analysis equation is:

$$(D^{1-\alpha})^{-A}(D^{1-\beta})^{B}(DQ^{-1})^{C}(D^{-\beta}Q)^{D} = 1$$
$$= (\omega_{01}Y/C_{00})^{-A}(\omega_{02}Y/C_{01})^{B}(\omega_{03}Y/C_{02})^{C}(\omega_{04}Y/C_{03})^{D}. \tag{18.24}$$

Since this is a maximization problem, the sign on the revenue term exponent is negative and the three cost terms is positive. The equations would be:

D terms	$-A(1-\alpha)$	$+B(1-\beta)$	$+C$	$-\beta D$	$= 0$	(18.25)
Q terms			$-C$	$+D$	$= 0$	(18.26)
Y Dual	$-A$	$+B$	$+C$	$+D$	$= 0$	(18.27)

From Equation (18.26) one observes that

$$C = D. \tag{18.28}$$

If one multiplies Equation (18.27) by $(1-\alpha)$ and subtracts Equation (18.25) and using Equation (18.28) one obtains:

$$B = [(1+\beta-2\alpha)/(\alpha-\beta)]D. \tag{18.29}$$

Using Equation (18.27) one obtains that

$$A = B + C + D.$$

Using the values for B and C in terms of D the result is:

$$A = [(1-\beta)/(\alpha-\beta)]D . \tag{18.30}$$

If one selects the value of $D = 1$, the values are $A = [(1-\beta)/(\alpha-\beta)]$, $B = [(1+\beta-2\alpha)/(\alpha-\beta)]$ and $C = 1$. Now Equation (18.24) becomes:

$$(\omega_{01}Y/C_{00})^{-[(1-\beta)/(\alpha-\beta)]} (\omega_{02}Y/C_{01})^{[(1+\beta-2\alpha)/(\alpha-\beta)]}$$
$$(\omega_{03}Y/C_{02})(\omega_{04}Y/C_{03}) = 1 . \tag{18.31}$$

Separating the constants and dual variables, the relationship obtained was:

$$\omega_{01}^{-[(1-\beta)/(\alpha-\beta)]} \omega_{02}^{[(1+\beta-2\alpha)/(\alpha-\beta)]} \omega_{03}\omega_{04}$$
$$= C_{00}^{-[(1-\beta)/(\alpha-\beta)]} C_{01}^{[(1+\beta-2\alpha)/(\alpha-\beta)]} C_{02}C_{03} = K . \tag{18.32}$$

The right-hand side of the equation consists of only constants, so the product is also a constant. If one substitutes the dual variables in terms of ω_{04}, the expression becomes quite complex and would need to be solved with specific values of the parameters. The equation would be:

$$[((1-\beta)/(\alpha-\beta))(1+\omega_{04})]^{-[(1-\beta)/(\alpha-\beta)]} \quad [(1-\alpha)/(\alpha-\beta)$$
$$+ (1+\beta-2\alpha)/(\alpha-\beta)\omega_{04}]^{[(1+\beta-2\alpha)/(\alpha-\beta)]} \quad \omega_{04}\omega_{04} = K . \tag{18.33}$$

18.3 EXAMPLE

The input parameters for the example are in Table 18.2.

Table 18.2: Parameters for Illustrative Problem		
Symbol	Value	Description of Variable or Parameter
A	$ 10	Set-up Cost ($/Batch)
i	0.10	Inventory Holding Cost Rate (Decimal Percent/Unit Time)
a	200	Scaling Constant for Price (Initial Price)
b	20	Scaling Constant for Cost (Initial Cost)
α	0.5	Price Elasticity with respect to Demand
β	0.1	Cost Elasticity with respect to Demand (Degree of Economies of Scale)
C_{00}	200	$C_{00} = a$
C_{01}	20	$C_{01} = b$
C_{02}	10	$C_{02} = A$
C_{03}	1	$C_{03} = ib/2$

Using the values of Table 18.2 in Equation (18.33), the result is:

$$[2.25(1+\omega_{04})]^{-2.25}[1.25+0.25\omega_{04}]^{0.25}\omega_{04}^2 = 1.4058 \times 10^{-4} \tag{18.34}$$

18. PROFIT MAXIMIZATION

Solving (18.34) via search techniques for ω_{04} the result is:

$$\omega_{04} = 0.0296 \tag{18.35}$$

Using this value for ω_{04}, the values of the remaining dual variables can be found from Equation (18.15), (18.17), and (18.19) as:

$$\omega_{03} = \omega_{04} = 0.0296 \tag{18.36}$$
$$\omega_{02} = [(1-\alpha)/(\alpha-\beta) + ((1+\beta-2\alpha)/(\alpha-\beta))\omega_{04}]$$
$$= [1.25 + 0.25 * (0.0296] = 1.2574 \tag{18.37}$$
$$\omega_{01} = [((1-\beta)/(\alpha-\beta))(1+\omega_{04})] = 2.25 * (1+0.0296) = 2.3166. \tag{18.38}$$

The value of the dual objective function is found from Equation (18.11) as:

$$D(Y) = -1[(C_{00}/\omega_{01})^{\omega}_{01}(C_{01}/\omega_{02})^{\omega}_{02}(C_{02}/\omega_{03})^{\omega}_{03}(C_{03}/\omega_{04})^{\omega}_{04}]^{-1} \tag{18.11}$$
$$= -[(200/2.3166)^{-2.3166}(20/1.2574)^{1.2574}(10/0.0296)^{0.0296}(1/0.0296)^{0.0296}]^{-1}$$
$$= -[0.0013982]^{-1}$$
$$= -715.2. \tag{18.39}$$

The minus sign indicates that it is a profit and not a cost. It is also interesting to note from the dual variables that the first cost term is the dominant cost term and that the second and third terms have equal value. The difference between the revenue dual variable and the sum of the cost dual variables is 1.0, which is the total of the dual variables when a cost only model is used.

The primal variables can be obtained from the dual variables using the primal-dual relationships. Rearranging the primal-dual relationship of Equation (18.20), the demand D can be evaluated as:

$$D = (\omega_{01}Y/C_{00})^{(1/(1-\alpha))} \tag{18.40}$$
$$= (2.3166 * 715.18/200)^{(1/0.5)}$$
$$= 68.6. \tag{18.41}$$

The value for the order quantity Q can be determined rearranging Equation (18.22) as:

$$Q = ((C_{02}D)/(\omega_{03}Y)) \tag{18.42}$$
$$= ((10 * 68.6)/(0.0296 * 715.18))$$
$$= 32.4. \tag{18.43}$$

The value of the primal objective function can now be found using Equation (18.10)

$$\text{Min } Y = -C_{00}D^{1-\alpha} + C_{01}D^{1-\beta} + C_{02}DQ^{-1} + C_{03}QD^{-\beta}. \tag{18.10}$$
$$= -200(68.6)^{0.5} + 20(68.6)^{0.9} + 10 * 68.6/32.4 + 1 * 32.4(68.2)^{-0.1}$$
$$= -1656.5 + 898.9 + 21.2 + 21.2$$
$$= -715.2. \tag{18.44}$$

The values of the primal and dual are in agreement. This problem was also solved by Yi Fang [2] via the transformed dual approach in which a constraint is added and the problem is solved via the constrained derivative approach as the degree of difficulty is one. The results are identical, although the solution approaches are quite different as the method presented obtains the additional equation via the method of dimensional analysis and solves a problem with zero degrees of difficulty.

18.4 TRANSFORMED DUAL APPROACH

The objective function of the primal is given by Equation (18.8) and is repeated here:

$$\text{Max } (\pi) = aD^{1-\alpha} - (bD^{1-\beta} + ADQ^{-1} + (ib/2)QD^{-\beta}). \tag{18.8}$$

Since the objective function was a signomial problem, Jung and Klein [1] used the transformed dual method developed by Duffin et al. [3] to solve the problem and their solution is presented. The problem transforms the primal objective function into a constraint with positive terms. The problem was formulated as:

$$\text{Max } z \tag{18.45}$$

subject to:

$$aD^{1-\alpha} - bD^{1-\beta} - ADQ^{-1} - (ib/2)QD^{-\beta} \geq z \tag{18.46}$$

The transformed primal problem is:

$$\text{Min } z^{-1} \tag{18.47}$$

subject to:

$$a^{-1}D^{\alpha-1}z + a^{-1}bD^{\alpha-\beta} + a^{-1}AD^{\alpha}Q^{-1} + a^{-1}(ib/2)QD^{\alpha-\beta-1} \leq 1 \tag{18.48}$$

The transformed primal function is a constrained posynomial function. The degree of difficulty remains the same as one variable and one terms was added, that is:

$$D = T - (N+1) = 5 - (3+1) = 1$$

From the coefficients and signs, the signum functions for the dual are:

$$\sigma_{01} = 1$$
$$\sigma_{11} = \sigma_{12} = \sigma_{13} = \sigma_{14} = 1$$
$$\sigma_1 = 1.$$

The dual problem formulation is:

Objective Function	ω_{01}		$= 1$	(18.49)
z terms	$-\omega_{01} +$	ω_{11}	$= 0$	(18.50)
D terms		$+(\alpha-1)\omega_{11} + (\alpha-\beta)\omega_{12} + \alpha\omega_{13} + (\alpha-\beta-1)\omega_{14} = 0$		(18.51)
Q terms		$-\omega_{13} + \omega_{14} = 0$		(18.52)

18. PROFIT MAXIMIZATION

Using Equations (18.49)–(18.52) the values of the dual variables ω_{01} and ω_{11} can be found and ω_{12} and ω_{13} can be found in terms of ω_{14}. The results are:

$$\omega_{01} = 1 \tag{18.53}$$
$$\omega_{11} = 1 \tag{18.54}$$
$$\omega_{12} = [(1 - \alpha + (1 + \beta - 2\alpha)\omega_{14})/(\alpha - \beta)] \tag{18.55}$$
$$\omega_{13} = \omega_{14} \tag{18.56}$$

and $\quad \omega_{10} = \omega_{11} + \omega_{12} + \omega_{13} + \omega_{14}$

or $\quad \omega_{10} = 1 + [(1 - \alpha - (2\alpha - \beta - 1)\omega_{14})/(\alpha - \beta)] + 2\omega_{14}$
$$= ((1 - \beta)(1 + \omega_{14}))/(\alpha - \beta)) \tag{18.57}$$

The dual objective function can be stated in terms of ω_{14} as:

$$D(Y) = [(1/\omega_{01})^{\omega_{01}} (a^{-1}\omega_{10}/\omega_{11})^{\omega_{11}} (a^{-1}b\omega_{10}/\omega_{12})^{\omega_{12}}$$
$$(a^{-1}A\omega_{10}/\omega_{13})^{\omega_{13}} ((a^{-1}ib/2)\omega_{10}/\omega_{14})^{\omega_{14}}]^{-1} \tag{18.58}$$

which, upon substitution, becomes:

$$D(Y) = [1]^1 * [a^{-1} * ((1 - \beta)(1 + \omega_{14}))/(\alpha - \beta)]^1 *$$
$$[[(a^{-1}b) * ((1 - \beta)(1 + \omega_{14}))/(\alpha - \beta)]/\{(1 - \alpha + (1 + \beta - 2\alpha)$$
$$\omega_{14}/(\alpha - \beta)\}]^{\{(1-\alpha+(1+\beta-2\alpha)\omega_{14}/(\alpha-\beta)\}}]$$
$$* [(a^{-1}A * ((1 - \beta)(1 + \omega_{14})/(\omega_{14}(\alpha - \beta))^{\omega_{14}}]$$
$$* [(a^{-1}(ib/2) * ((1 - \beta)(1 + \omega_{14})/(\omega_{14}(\alpha - \beta))^{\omega_{14}}] \tag{18.59}$$

The logarithm form of Equation (18.59) is:

$$\text{Log}[D(Y)] = \log[(a^{-1}(1 + \beta)/(\alpha - \beta)) * (1 + \omega_{14})]$$
$$+ \{(1 - \alpha + (1 + \beta - 2\alpha)\omega_{14}/(\alpha - \beta)\}) \log[[(a^{-1}b)$$
$$* ((1 - \beta)(1 + \omega_{14}))/(\alpha - \beta)/\{(1 - \alpha + (1 + \beta - 2\alpha)\omega_{14}/(\alpha - \beta))\}]$$
$$+ \omega_{14} \log(a^{-1}A * (1 - \beta)(1 + \omega_{14}))/(\omega_{14}(\alpha - \beta))$$
$$+ \omega_{14} \log(a^{-1}(ib/2) * (1 - \beta)(1 + \omega_{14}))/(\omega_{14}(\alpha - \beta)) \,. \tag{18.60}$$

Taking the derivative of Equation (18.60) and setting it to zero and reducing the terms results in:

$$\text{Log}\{[(a^{-1}b)(1 - \beta)(1 + \omega_{14})]/[1 - \alpha + (1 + \beta - 2\alpha)\omega_{14}]\}^{(1+\beta-2\alpha))/(\alpha-\beta)}$$
$$+ \text{Log}[a^{-1}A((1 - \beta)/(\alpha - \beta)) * ((1 + \omega_{14})/\omega_{14})]$$
$$+ \text{Log}[a^{-1}(ib/2)((1 - \beta)/(\alpha - \beta)) * ((1 + \omega_{14})/\omega_{14})] = 0 \tag{18.61}$$

Taking the anti-log this becomes:

$$\{[(a^{-1}b)(1 - \beta)(1 + \omega_{14})]/[1 - \alpha + (1 + \beta - 2\alpha)\omega_{14}]\}^{(1+\beta-2\alpha))/(\alpha-\beta)}$$
$$* [a^{-1}A((1 - \beta)/(\alpha - \beta)) * ((1 + \omega_{14})/\omega_{14})]$$
$$* [a^{-1}(ib/2)((1 - \beta)/(\alpha - \beta)) * ((1 + \omega_{14})/\omega_{14})] = 1 \tag{18.62}$$

18.4. TRANSFORMED DUAL APPROACH

Using the values for the example problem of $\alpha = 0.5, \beta = 0.1, a = 200, b = 20, A = 10$, and $i = 0.10$ to solve for ω_{14}, Equation (18.62) becomes;

$$\{[(200^{-1}20)(1-0.1)(1+\omega_{14})]/[1-0.5+(1+0.1-2*0.5)*\omega_{14}]\}^{[(1+0.1-2*0.5)/(0.5-0.1)]}$$
$$*[(200^{-1}*10*((1-0.1)/(0.5-0.1))*((1+\omega_{14})/\omega_{14})]$$
$$*[(200^{-1}*(0.1*20/2)*((1-0.1)/(0.5-0.1))*((1+\omega_{14})/\omega_{14})] = 1 \quad (18.63)$$

or

$$[0.09*(1+\omega_{14})/[0.5+0.1*\omega_{14}]^{0.25}$$
$$*[0.1125*(1+\omega_{14})/(\omega_{14})]*[0.01125*(1+\omega_{14})(\omega_{14}) = 1 \quad (18.64)$$

Solving for ω_{14} one obtains:

$$\omega_{14} = 0.0296 \quad (18.65)$$

This is the same result as obtained by the previously in Equation (18.35) for ω_{04} and thus the solutions will be the same as that of the dimensional analysis approach used. The constraint dual variables are the same as the objective dual variables as the constraint was equivalent to the earlier objective function. The remaining dual variables are:

$$\omega_{12} = 1.2574 \text{ (same as } \omega_{02} \text{ earlier,}$$
$$\text{but represents the same function in this problem)} \quad (18.65)$$
$$\omega_{13} = 0.0296 \text{ (same as } \omega_{03} \text{ earlier,}$$
$$\text{but represents the same function in this problem)} \quad (18.66)$$
$$\omega_{11} = 1.0 \quad (18.67)$$
$$\omega_{10} = 2.3167 \quad (18.68)$$

The value of the dual objective function using Equation (18.58) is

$$D(Y) = [(1/\omega_{01})^{\omega_{01}}(a^{-1}\omega_{10}/\omega_{11})^{\omega_{11}}(a^{-1}b\omega_{10}/\omega_{12})^{\omega_{12}}$$
$$(a^{-1}A\omega_{10}/\omega_{13})^{\omega_{13}}((a^{-1}ib/2)\omega_{10}/\omega_{14})^{\omega_{14}}]^{-1} \quad (18.58)$$
$$= [(1/1)^1 * (200^{-1}*2.3174/1)^1$$
$$*(200^{-1}*20*2.3167/1.2574)^{1.2574}$$
$$*(200^{-1}*10*2.3167/0.0296)^{0.0296}$$
$$*(200^{-1}*(0.1*20/2)*2.3167/0.0296)^{0.0296}]^{-1}$$
$$= [1.3978*10^{-3}]^{-1}$$
$$= 715.4 . \quad (18.69)$$

The dual solution is in agreement with the dual solution obtained using the dimensional analysis approach. The solution for the primal variables can be obtained from the primal-dual relationships. The equations used to solve for the dual variables were quite different, but gave the same value for ω_{14} and its equivalent of ω_{04} of 0.0286. The approach using dimensional analysis was somewhat easier as it did not require taking the derivative of the log of the dual and it had one less dual variable than the transformed dual approach.

18.5 EVALUATIVE QUESTIONS

1. Solve the example problem changing the price elasticity to determine the effects of D and Q and compare the results with the original example problem.

   Table 18.3: Parameters for Problem 1

Symbol	Value	Description of Variable or Parameter
A	$ 10	Set-up Cost ($/Batch)
i	0.10	Inventory Holding Cost Rate (Decimal Percent/Unit Time)
a	200	Scaling Constant for Price (Initial Price)
b	20	Scaling Constant for Cost (Initial Cost)
α	0.3	Price Elasticity with respect to Demand
β	0.1	Cost Elasticity with respect to Demand

2. Solve the example problem changing the set-up cost to determine the effects on D and Q and compare the results.

   Table 18.4: Parameters for Problem 2

Symbol	Value	Description of Variable or Parameter
A	$ 80	Set-up Cost ($/Batch)
i	0.10	Inventory Holding Cost Rate (Decimal Percent/Unit Time)
a	200	Scaling Constant for Price (Initial Price)
b	20	Scaling Constant for Cost (Initial Cost)
α	0.5	Price Elasticity with respect to Demand
β	0.1	Cost Elasticity with respect to Demand

3. Solve for the primal variables using the dual variables obtained in the transformed dual approach example problem.

4. Take the derivative of Equation (18.59) and show all the steps to obtain Equation (18.63) including the terms that are canceled.

5. Discuss the advantages and disadvantages of the methods of dimensional analysis versus the constrained derivative approach.

REFERENCES

[1] Jung, H., Klein, C. M., Optimal inventory policies under decreasing cost functions via geometric programming, *European Journal of Operational Research* 132 (2001), 628–642. 103, 109

[2] Yi Fang, IENG 593 F, "Geometric Programming Research Paper," April 15, 2010, 23 pages. 109

[3] Duffin, R. J., Peterson, E. L., Zener, C., *Geometric Programming – Theory and Application*, Wiley, New York, 1976. 109

CHAPTER 19

Summary and Future Directions

19.1 SUMMARY

The object of this text is to generate interest in geometric programming amongst manufacturing engineers, design engineers, manufacturing technologists, cost engineers, project managers, industrial consultants and finance managers by illustrating the procedure for solving certain industrial and practical problems. The various case studies were selected to illustrate a variety of applications as well as a set of different types of problems from diverse fields. Several additional problems were added focusing on profit maximization and additional problems with degrees of difficulty. In addition, the methods of dimensional analysis and the constrained derivative approach have been presented in detail. Table 19.1 is a summary of the case studies presented in this text giving the type of problem, degrees of difficulty, and other details.

The metal removal economics example also had variable exponents in the general solution. The problems were worked in detail so general solutions could be obtained and also to show that the dual and primal solutions were identical. The problems were selected to illustrate a variety of types and also to show the use of the primal-dual relationships to determine the equations for the primal variables. It is by showing the various types of applications in detailed examples that others can follow the procedure and develop new applications.

19.2 FUTURE DIRECTIONS

The author is hopeful that others will communicate him additional examples to illustrate new applications that can be included in future editions. New applications will attract new practitioners to this fascinating area of geometric programming. It is believed that the scope of geometric programming will expand with new applications.

The author would like to include some software for different applications in geometric programming in the future and would welcome contributions.

19.3 DEVELOPMENT OF NEW DESIGN RELATIONSHIPS

There are many different types of problems that can be solved by geometric programming and one of the significant advantages of the method is that it is possible in many applications to develop general design relationships. The general design relationships can save considerable time and effort in instances where the constants are changed.

19. SUMMARY AND FUTURE DIRECTIONS

Although geometric programming was first presented nearly 50 years ago, the applications have been rather sparse compared to that of linear programming. One goal is that as researchers take advantage of the potential to develop design relationships that new applications will rapidly occur. The development of new design relationships can significantly reduce the development time and cost for new products and this is essential for companies to remain competitive in the global economy.

19.3. DEVELOPMENT OF NEW DESIGN RELATIONSHIPS

Table 19.1: Summary of Case Study Problem

Chapter	Case Study	Degrees of Difficulty	Number of Constraints	Number of Variables	Variable Description	Number of Solutions	Special Characteristics
4	The Optimal Box Design	0	1	3	Height, Width, Length	1	
5	Trash Can	0	1	2	Height, Diameter	1	
6	Open Cargo Shipping Box	0	1	3	Height, Width, Length	1	
7	Metal Casting Cylindrical Riser	0	1	2	Height, Diameter	1	Classical Problem
8	Inventory Model	0	0	1	Lot Size	1	
9	Process Furnace Design	0	1	3	Temperature, Length, Height	1	Dominant Equation Negative Dual Variable Dominant Equation
10	Gas Transmission Pipe Line	0	1	4	Length, Diameter, Flow Length Pressure Ratio Factor	1	Four Variables
11	Profit Maximization	0	0	3		1	Cobb-Douglas Profit Function
12	Material Removal/Metal Cutting	0	1	2	Feed Rate, Cutting Speed	1	Fractional Exponents
13	Journal Bearing Design	1	1	2	Journal Radius, Bearing Half-Length	1	Dimensional Analysis Approach
14	Metal Casting Hemispherical Top Riser	2	1	2	Height, Diameter	1	Dimensional Analysis Approach
15	Liquefied Petroleum(LPG) Cylinder	1	1	2	Height, Diameter	2	Multiple Solutions Dimensional Analysis Approach
16	Metal Cutting Economics-2 Constraints	1	2	2	Feed Rate, Cutting Speed	3	Multiple Solutions Derivative Approach
17	Open Cargo Shipping Box with Skids	1	0	3	Height, Width, Length	1	Derivative Approach Dimensional Analysis Approach Condensation of Terms Approach
18	Profit Maximization with Decreasing Cost Functions	1	1	2	Lot Size, Demand	1	Dimensional Analysis Approach Transformed Dual Approach

CHAPTER 20

Thesis and Dissertations on Geometric Programming

A search was made to find the thesis and dissertations on geometric programming. The listed items had geometric programming listed in the abstract, but that does not necessarily imply that the research focused on geometric programming. However, most the research work did focus on geometric programming. The search was made using Dissertation Abstracts Online and they do have the thesis/dissertation available for sale on their web site.

20. THESIS AND DISSERTATIONS ON GEOMETRIC PROGRAMMING

	Author	Title	School	Year	Type
1	Cheung, Wing Tai	Geometric Programming and Signal Flow Graph Assisted Design of Interconnect and Analog Circuits	University of Hong Kong	2008	M. Phil
2	Firouzabadi, Sina	Jointly Optimal Placement and Power Allocation of Nodes of a Wireless Network	University of Maryland	2007	MS
3	Gupta, Deepak Prakash	Energy-Sensitive Machining Parameter Optimization Model	West Virginia University	2005	MSIE
4	Kokatnur, Ameet	Design and Tevelopment of a Target-Costing Model for Machining	West Virginia University	2004	MSIE
5	Helba, Michael JoF n	Dev. and Opt. of a NL Multibumper Design for Spacecraft Protective Structures Using Simulated Annealing	Univ. of Alabama, Huntsville	1994	MS
6	Liu, Guili	The Hoelder Inequality and Its Application in Banach Algebra and Geometric Programming	The University of Regina(CAN)	1992	MSc
7	Yang, Edward V.	Sensitivity of Optimum Heat Exchanger Design by Geometric Programming	Univ. of Massachusetts-Lowell	1976	MS
8	Meteer, James W.	A Geometric Programming Solution to a Strata Allocation Problem	University of Wyoming	1976	MS
9	Yeh, Lucia Lung-Chia	Geometric Programming Solution of a Problem Involving Sampling from Overlapping Frames	University of Wyoming	1974	MS
10	Shah, Shivji Virji	Optimization of Finned Tube Condenser by Geometric Programming	Univ. of Massachusetts-Lowell	1973	MS
11	Shah, Hemendrakurrar R.	Optimal Heat-Exchanger Design by Geometric Programming	Univ. of Massachusetts-Lowell	1971	MS

Figure 20.1: Original data was obtained from Dissertation Abstracts Online.

	Author	Title	School	Year	Type
1	Charafeddine, Mohamad	Communication over n-user Interference Channel	Stanford University	2009	Ph.D.
2	Kim, Jintae	Multi-level Design Optimizations of Pipelined A/D Converter	UCLA	2008	Ph.D.
3	Liu, Hongbo	Cross-Layer Design for Reliable and Efficient Data Transmission over Multiple Antenna Mobile Infostation Networks	Rutgers University	2008	Ph.D.
4	Tan, Chee Wei	Nonconvex Power Control in Multiuser Communication Systems	Princeton University	2008	Ph.D.
5	Joshi, Siddarth	Large-Scale Geometric Programming for Devices and Circuits	Stanford University	2008	Ph.D.
6	Roy, Sanghamitra	Novel Modeling and Optimization Techniques for Nano-scale VLSI Designs	Wisconsin University	2008	Ph.D.
7	Patil, Dinesh	Design of Robust Energy-Efficient Digital Circuits Using Geometric Programming	Stanford University	2008	Ph.D.
8	Seong, Kibeom	Cross-Layer Resource Allocation for Multi-user Communication Systems	Stanford University	2008	Ph.D.
9	Hsiung, Kan-Lin	Geometric Programming Under Uncertainty with Engineering Applications	Stanford University	2008	Ph.D.
10	Zheng, Gan	Optimization in Linear Multiuser MIMO Systems	University of Hong Kong	2007	Ph.D.
11	Zhang, Wei	Bivariate Cubic L1 Splines and Applications	North Carolina State	2007	Ph.D.
12	Zhou, Quming	Reliability-driven Circuit Optimization and Design	Rice University	2007	Ph.D.
13	Chinnery, David Graeme	Low Power Design Amtomation	UC-Berkeley	2006	Ph.D.
14	Colleran, David M.	Optimization of Phase-locked Loop Circuits via Geometric Programming	Stanford University	2006	Ph.D.
15	Vanderhaegen, John Peter	A Design Methodology for Analog Circuits Based on Global Optimization	UC-Berkeley	2005	Ph.D.
16	Wang, Yong	Theory and Algorithms for Shape-preserving Bivariate L1 Splines	North Carolina State	2005	Ph.D.
17	Yun, Sunghee	Convex Optimization for Digital Integrated Circuit Applications	Stanford University	2005	Ph.D.
18	Xu, Yang	Affordable Analog and Radio Frequency Integrated Circuits Design and Optimization	Carnegie Mellon University	2004	Ph.D.
19	O'Neill, Daniel C.	Jointly Optimal Network Performance: A Cross-layer Approach	Stanford University	2004	Ph.D.
20	Chiang, Mung	Solving Nonlinear Problems in Communication Systems Using Geometric Programming and Dualities	Stanford University	2003	Ph.D.
21	Qin, Zhanhai	Topological Circuit Reduction: Theory and Applications	UC-San Diego	2003	Ph.D.
22	Cheng, Hao	Tehory and Algorithms for Cubic L(1) Splines	North Carolina State	2002	Ph.D.
23	Jacobs, Etienne Theodorus	Power Dissapation and Timing in CMOS Circuits	Technische Univ. Eindhoven	2001	Dr.
24	Jung, Hoon	Optimal Inventory Policies for an Economic Order Quantity Models Under Various Cost Functions	Univ. Missouri-Columbia	2001	Ph.D.
25	Hershenson, Maria del Mar	CMOS Analog Circuit Design via Geometric Programming	Stanford University	2000	Ph.D.
26	Lou, Jinan	Integrated Logical and Physical Optimizations for Deep Submicron Circuits	Univ. Southern California	1999	Ph.D.
27	Xu, Li Na	Optimization Methods for Computing Empirically Constrained Extremal Probability Distributions	The University of Iowa	1996	Ph.D.
28	Zao, John Kar-kin	Finite-Precision Representation and Data Abstraction for Three-Dimensional Euclidean Transformations	Harvard University	1995	Ph.D.
29	Maranas, Costas D.	Global Optimization in Computational Chemistry and Process Systems Engineering (Potential Energy)	Princeton University	1995	Ph.D.
30	Ince, Erdem	A Parallel Balanced Posynomial Unconstrained Geometric Programming Algorithm	Colorado School of Mines	1995	Ph.D.

Figure 20.1: Thesis and Dissertations on Geometric Programming. *Continues.*

20. THESIS AND DISSERTATIONS ON GEOMETRIC PROGRAMMING

	Author	Title	School	Year	Type
31	Yang, Hsu-Hao	Investigation of Path-Following Algorithms for Signomial Geometric Programming Problems	The University of Iowa	1994	Ph.D.
32	Jackson, Jack Allen, Jr.	A Mathematical Experiment in Dual Geometric Programming	Colorado School of Mines	1994	Ph.D.
33	Mader, Douglas Paul	An Improved Method for Sensitivity Analysis in Geometric Programming	Colorado School of Mines	1994	Ph.D.
34	Cheng, Zhao-Yang	A Least Squares Approach to Interior Point Methods with an Application to Geometric Programming	Rensselaer Poly. Institute	1993	Ph.D.
35	Zhu, Jishan	On the Path-Following Methods for Linearly constrained Convex Programming Problems	The University of Iowa	1992	Ph.D.
36	Narang, Ramesh Vashoomal	Issues and Methodologies in Auto. Proc. Planning for a Three-Dimensional Part Representation to Recomend Machining Parameters	The University of Iowa	1992	Ph.D.
37	Martino, Thomas John	A Tandem Joint Inspection and Queueing Model Employing Posynomial Geometric Programming	University of Rhode Island	1991	Ph.D.
38	Katz, Joseph Harold	An Algorithm for Solving a Class of Nonlinear, Unconstrained, Multi-Variable, Signomial Optimization Problems (GP)	Colorado School of Mines	1991	Ph.D.
39	Dull, Owen S.	A G P Based Method for Determining the Required Constraints in a Class of Chemical Blending Problems	Colorado School of Mines	1991	Ph.D.
40	Viriththamulla, Ganage I.	Mathematical Programming Models and Heuristics for Standard Modular Design Problem	The University of Arizona	1991	Ph.D.
41	Knowles, James Alexander	A Geometric Programming Approach to the Solution of Multistage Countercurrent Heat Exchanger Systems	Colorado School of Mines	1990	Ph.D.
42	Rama, Dasaratha V	Planning Models for Communications Satellites Offering Customer Premises Services	The University of Iowa	1990	Ph.D.
43	No, Hoon	Accelerated Convergent Methods for a Class of Nonlinear Programming Problems	The University of Iowa	1990	Ph.D.
44	Wessels, Gysbert Johannes	A Geometric Programming Algorithm for Solving a Class of Nonlinear, Signomial Optimization Problems	Colorado School of Mines	1989	Ph.D.
45	Bailey, Steven Scott	An Algorithm for the Solution of a Class of Economic Models for the Design of Prestressed Concrete Bridges	Colorado School of Mines	1989	Ph.D.
46	Kirk, John B.	A Geometric Programming Based Algorithm for Preprocessing Nonlinear Signomial Optimization Programs	Colorado School of Mines	1988	Ph.D.
47	Thome, James John, Jr.	An Algorithm for Solving a Class of Nonlinear, Unconstrained, Signomial Economic Models Using the Greening Technique	Colorado School of Mines	1988	Ph.D.
48	Plenert, Gerhard Johannes	A Solution to Smith's General Case Bottleneck Problem Allowing for an Unlimited Number of Scheduled Jobs	Colorado School of Mines	1987	Ph.D.
49	Chibani, Lahbib	Integrated Optimal Design of Structures Subjected to Alternate Loads Using Geometric Programming	University of Illinois-U-C	1987	Ph.D.
50	Muhammed, Abdelfatah A.	Information Theory and Queueing Theory Via Generalized Geometric Programming	North Carolina State	1987	Ph.D.
51	Mrema, Godwill E. E. C.	Computation of Chemical and Phase Equilibria	Universitetet I Trondheim	1987	Dr. Ing
52	Li, Xing-Si	Entropy and Optimization	University of Liverpool	1987	Ph.D.
53	Choi, Jae Chul	Generalized Benders' Decomposition for Geometric Programming with Several Discrete Variables	The University of Iowa	1987	Ph.D.
54	Arreola Contreras, Jose J.	Semi-discrete Geometric Programming	University of Pittsburgh	1986	Ph.D.
55	Tsai, Pingfang	An Optimization Algorithm and Economic Analysis for a Constrained Machining Model	West Virginia University	1986	Ph.D.
56	Rajopal, Jayant	A Duality Theory for Geometric Programming Based on Generalized Linear Programming	The University of Iowa	1985	Ph.D.
57	Murphy, Robert Craig	A Dual-Dual Algorithm for the Solution of a Class of Nonlinear Resource Allocation Models	Colorado School of Mines	1985	Ph.D.
58	Burns, Scott Allen	Structural Optimization Using Geometric Programming and the Integrated Formulation	University of Illinois-U-C	1985	Ph.D.
59	Djanali, Supeno	Geometric Programming and Decomposition Techniques in Optimal Control	University of Wisconsin	1984	Ph.D.
60	Rajasekera, Jayantha R.	Perturbation Techniques for the Solution of Posynomial, Quadratic and LP Approximation Programs	North Carolina State University	1984	Ph.D.

Figure 20.1: *Continued.* Thesis and Dissertations on Geometric Programming. *Continues.*

	Author	Title	School	Year	Type
61	Tang, Che-Chung	Mathematical Models and Optimization Techniques for Analysis and Design of Wastewater Treatment Systems	University of Illinois-U-C	1984	Ph.D.
62	Ravanbakht, Camela	Geometric Programming: An Efficient Computer Algorithm for Traffic Assigner	North Carolina State University	1984	Ph.D
63	Chou, Jaw Huoy	Contributions to Nondifferential Mathematical Programming	North Carolina State University	1984	Ph.D
64	Kyparsis, Jerry	Sensitivity and Stability for Nonlinear and Geometric Programming: Theory and Applications	George Washington University	1983	D.SC.
65	Dziuban, Stephen T.	Ellipsoid Algorithm Variants in Nonlinear Programming	Rensselaer Poly. Tech.	1983	Ph.D
66	Morningstar, Kay Gladys	Self-Teaching Enrichment Modules on Modern Applications of Mathematics for Talented Secondary Students and Evaluation...	University of Delaware	1983	Ph.D
67	Duggar, Cynthia Miller	A Study of the Relationships Among Computer Programming Ability, Computer Program Content, Computer Programming Style...	Georgia State University	1983	Ph.D
68	Ohanomah, Matthew O.	Computational Algorithms for Multicomponent Phase Equilibria and Distillation	Univ. of British Columbia	1982	Ph.D
69	Marin Gracia, Angel	Contribuciones A La Teoria Matematica De La Planificacion Del Transporte Y Sus Aplicaciones	Univ. Politecnica De Madrid	1982	Ph.D
70	Mayer, Robert Hall, Jr	Cost Estimates from Stochastic Geometric Programs	University of Delaware	1982	Ph.D
71	Brar, Guri Singh	Geometric Programming for Engineering Design Optimization	The Ohio State University	1981	Ph.D
72	Memon, Altaf Ahmed	Environmental Noise Management: An Optimization Approach	University of Pittsburgh	1980	Ph.D
73	Silberman, Gabriel Mauricio	The Design and Evaluation of an Active Memory Unit	SUNY-Buffalo	1980	Ph.D
74	Junna, Mohan Reddy	Irrigation System Improvement by Simulation and Optimization	Colorado State University	1980	Ph.D
75	Rosenberg, Eric	Globally Convergent Algorithms for Convex Programming with Applications to Geometric Programming	Stanford University	1979	Ph.D
76	Hough, Clarence Lee, Jr.	Optimization of the Second-order Logarithmic Machining Economics Problem by Extended Geometric Programming	Texas A&M University	1978	Ph.D
77	Yale, Wilson Winant	The Design of Solid Waste Systems: An Application of Geo. Prog. to Problems in Municipal Solid Waste Management	Lehigh University	1978	Ph.D
78	Sheldon, Dan	Nonlinear Optimization by Geometric Programming Methods with Applications	UCLA	1978	Ph.D
79	Wong, Shek-Nam Danny	Nonlinear Models for Systems Planning and Control: Solution and Analyses Via Geometric Programming	The Pennsylvania State Univ.	1978	Ph.D
80	Maurer, Ruth Allene Lamb	A Geometric Programming Approach to the Preliminary Design of Wastewater Treatment Plants	Colorado School of Mine	1978	Ph.D
81	Ramamurthy, Subramanian	Structural Optimization Using Geometric Programming	Cornell University	1977	Ph.D
82	Grange, Franklin Earnest, II	The Solution and Analysis of the Geometric Mean Portfolio Problem with Methods of Geometric Programming	Colorado School of Mine	1977	Ph.D
83	Kise, Peter Ethan	Systems of Optimization via Geometric Programming	University of Pennsylvania	1976	Ph.D
84	Wu, Chin Chang	Design and Modeling of Solar Sea Power Plants by Geometric Programming	Carnegie-Mellon University	1976	Ph.D
85	Gribik, Paul Robert	Semi-infinite Programming Equivalents and Solution Techniques for Optimal Experimental Design and GP Problems...	Carnegie-Mellon University	1976	Ph.D
86	McMasters, John Hamilton	The Optimization of Low-speed Flying Devices by Geometric Programming	Purdue University	1975	Ph.D
87	Hall, Michael Anthony	A Geometric Programming Approach to Highway Network Equilibria	Northwestern University	1974	Ph.D
88	Emery, Sidney Williams, Jr.	Transoceanic Dry-Bulk Systems Optimization by Geometric Programming	Stanford University	1974	Ph.D
89	Piecnik, Joseph Matthew	Optimization of Engineering Systems using Extended Geometric Programming	The Ohio State University	1974	Ph.D
90	Zorachi, Michael John	A Primal Method for Geometric Programming	Rensselaer Polytechnic Institute	1974	Ph.D

Figure 20.1: *Continued.* Thesis and Dissertations on Geometric Programming. *Continues.*

20. THESIS AND DISSERTATIONS ON GEOMETRIC PROGRAMMING

	Author	Title	School	Year	Type
91	Salinas-Pacheco, Juan Jose	Minimum Cost Design of Concrete Beams Using Geometric Programming	University of Calgary	1974	Ph.D
92	Smeers, Yves Marie	Geometric Programming with Applications to Management Science	Carnegie-Mellon University	1973	Ph.D
93	Wiebking, Rolf D	Deterministic and Stochastic G P Models for Optimal Engineering Design in Electrical Power Generation & Simulation	Rensselaer Poly. Institute	1973	Ph.D
94	Ben-Tal, Aharon	Contributions to Geometric Programming and Generalized Convexity	Northwestern University	1973	Ph.D
95	Vandrey, Joan Betty W.	Reversed Generalized Geometric Programming	Northwestern University	1973	Ph.D
96	Planchart, Alejo	A Geometric Programming Approach to the Problem of Finding Efficient Algorithms for Constrained Location Prob.	Northwestern University	1973	Ph.D
97	McCarl, Bruce Alan	A Computational Study of Polynomial Geometric Programming	The Pennsylvania State Univ.	1973	Ph.D
98	Mekaru, Mark Masanobu	N-Dimensional Transportation Problems: Algorithm for Linear Problems and Application of GP for non-linear Probs.	Arizona State University	1973	Ph.D
99	Jefferson, Thomas Richard	Geometric Programming with an Application to Transportation Planning	Northwestern University	1972	Ph.D
100	Beck, Paul Alan	A Modified Convex Simplex Algorithm for Geometric Programming with Subsidiary Problems	Rensselaer Poly. Institute	1972	Ph.D
101	Dinkel, John Joseph	A Duality Theory for Dynamic Programming Problems via Geometric Programming	Northwestern University	1971	Ph.D
102	Unklesbay, Kenneth Boyd	The Extension of Geometric Programming to Optimal Mechanical Design	Univ. of Missouri-Columbia	1971	Ph.D
103	Oleson, Gary K.	Computational Aspects of Geometric Programming	Iowa State University	1971	Ph.D
104	McNamara, John Richard	The Optimal Design of Water Quality Management Systems: An Application of Multistage Geometric Programming	Rensselaer Poly. Institute	1971	Ph.D
105	Staats, Glenn Edwin	Computational Aspects of Geometric Programming with Degrees of Difficulty	University of Texas-Austin	1970	Ph.D
106	Teske, Clarence Eugeae	Optimal Structural Design Using Geometric Programming	Texas Tech University	1970	Ph.D
107	Kochenberger, Gary Austin	Geometric Programming: Extensions to Deal with Degrees of Difficulty and Loose Constraints	Univ. of Colorado-Bolder	1969	D.B.A.
108	Pascual, Luis De La Cruz	Constrained Maximization of Posynomials and Vector-Valued Criteria in Goemetric Programming	Northwestern University	1969	Ph.D
109	Ecker, Joseph George	Geometric Programming: Duality in Quadratic Programming and LP Approximation	The University of Michigan	1968	Ph.D
110	Passy, Ury	Generalization of Geometric Programming: Partial Control of Linear Inventory Systems	Stanford University	1966	Ph.D
111	Avriel, Mordecai	Topics in Optimization: Block Search; Applied and Stochastic Geometric Programming	Stanford University	1966	Ph.D

Figure 20.1: *Continued.* Thesis and Dissertations on Geometric Programming.

Author's Biography

DR. ROBERT C. CREESE

Dr. Robert C. Creese, Certified Cost Engineer (CCE), is Professor of Industrial and Management Systems Engineering at West Virginia University, USA and recently has taught courses on Engineering Economy, Advanced Engineering Economics, Cost and Estimating for Manufacturing, Manufacturing Processes and Advanced Manufacturing Processes. He has previously taught at The Pennsylvania State University (9 years), Grove City College (4 years), Aalborg University in Denmark (3 sabbaticals) and at West Virginia University for over 31 years.

He is a Fellow of the Association for the Advancement of Cost Engineering, International (AACEI), received the Charles V. Keane Service Award and Brian D. Dunfield Educational Service Award presented by AACE, and has been treasurer of the Northern West Virginia Section of AACE for more than 20 years. He is a Life Member of AACE International, ASEE (American Society for Engineering Education) and ASM (American Society for Materials). He also is a member of ISPA, SCEA, AIST, AWS, and AFS.

He obtained his B.S. Degree in Industrial Engineering from the Pennsylvania State University, his M.S. Degree in Industrial Engineering from the University of California at Berkeley, and his Ph.D. Degree in Metallurgy from the Pennsylvania State University.

He has authored the book *Introduction to Manufacturing Processes and Materials* (Marcel Dekker-1999) and co-authored two books *Estimating and Costing for the Metal Manufacturing Industries* (Marcel Dekker-1992) with Dr. M. Adithan of VIT University Vellore, India and Dr. B. S. Pabla of the Technical Teachers' Training Institute, Chandigarh, India and *Strategic Cost Analysis for Project Managers and Engineers* (New Age International Publishers-2010) with Dr. M. Adithan, VIT University, Vellore, India. He has authored and co-authored more than 100 technical papers.

Index

Arithmetic Mean, 4

Box design problem, 13

Cargo Shipping Box, 21, 95
Chvorinov's Rule, 27, 67
Classical Problem, 95
Cobb-Douglas Production Function, 47, 49
Compressor Pressure Ratio Factor, 43
Condensation of Terms Approach, 95
Constrained Derivative Approach, 64, 99, 102, 109, 112, 115
Cost Functions, 36, 112
Cylindrical Riser Design, 29, 30

Derivative Approach, 64, 95–97, 99, 102, 109, 112, 115
Design of LPG (Propane gas) Cylinder Design (Propane Gas) Cylinder(LPG) design, 77
Design of LPG (Propane gas) Cylinder Design (Propane Gas) Cylinder(LPG) design Cylinder, 77
Design of LPG (Propane Gas) Cylinder(LPG) design, 77
Design of LPG (Propane Gas) Cylinder(LPG) design Cylinder, 77
Design Relationships, 3, 24, 31, 75, 115, 116
Dimensional Analysis Technique, 83
Dual Objective Function, 8, 47, 48, 64, 72, 81, 82, 88, 90, 101, 105, 108, 110, 111

Dual Variables, 8, 9, 14, 18, 20, 22, 23, 29, 34, 36, 38–40, 44, 46, 48, 49, 60–62, 64, 65, 69–71, 80, 81, 88–91, 95, 100, 105, 107, 108, 110–112

Furnace Design, 37

Gas Transmission (Pipeline) design, 43
Geometric Mean, 4
Geometric Programming, 3–7, 10, 13, 16, 17, 21, 25, 27, 30, 31, 33, 36, 37, 47, 50, 55, 65, 75, 77, 87, 92, 95, 112, 115, 116, 119, 121–124
  Advantages, 115
  Applications, 17, 115, 116
  Engineering Design, 5, 6, 10
  History, 4, 5
  Pioneers, 5

Inequality Constraints, 8, 9

Journal Bearing Design, 59

Linear Programming, 4, 9
LPG (Propane gas) Cylinder Design, 77

Machining Economics, 55, 92
Material Removal Economics, 51
Metal Casting, 5, 27, 31
Metal cutting (material removal) Economics, 51, 54, 55, 85, 86, 91, 92, 102
Metal cutting (Metal cutting (material removal) Economics) Economics, 51, 55, 85, 86, 91, 92, 102

INDEX

Multiple Solutions, 85

Non-negativity Conditions, 8

Open Cargo Shipping Box, 21, 95
Optimal Box Design Problem, 13
Optimization, 3–6, 10, 41, 55, 92
    design optimization, 3
    techniques of optimization (methods), 3
Orthogonal Conditions, 8

Primal and Dual Formulation, 54
Primal Objective Function, 8, 13, 17, 22, 47, 96, 102, 105, 108, 109

Profit Maximization, 47, 103, 115
Propane Gas Cylinder(LPG) design, 77

Riser Design, 5, 6, 27, 29–31, 67, 68, 75
    Cylindrical Riser, 29–31, 75

Scale of Production, 47
Search Techniques, 108
Summary, 115

Techniques of Optimization Methods, 3
Transformed Dual Approach, 109, 111, 112
Trash Can, 17, 20

Lightning Source UK Ltd.
Milton Keynes UK
UKOW07f0751030816

279820UK00003B/231/P